Doing Data Analysis with Minitab 12

ROBERT CARVER

DUXBURY PRESS
An Imprint of Brooks/Cole Publishing Company
I(T)P® An International Thomson Publishing Company

Pacific Grove • Albany • Belmont • Bonn • Boston • Cincinnati • Detroit
Johannesburg • London • Madrid • Melbourne • Mexico City • New York
Paris • Singapore • Tokyo • Toronto • Washington

Assistant Editor: *Cindy Mazow*
Editorial Assistant: *Rita Jaramillo*
Production Editor: *Mary Vezilich*
Printing and Binding: *Patterson Printing*

For more information, contact Duxbury Press at Brooks/Cole Publishing Company, or electronically at **http://www.duxbury.com**

BROOKS/COLE PUBLISHING COMPANY
511 Forest Lodge Road
Pacific Grove, CA 93950
USA

International Thomson Editores
Seneca 53
Col. Polanco
11560 México, D. F., México

International Thomson Publishing Europe
Berkshire House 168-173
High Holborn
London WC1V 7AA
England

International Thomson Publishing Japan
Hirakawacho Kyowa Building, 3F
2-2-1 Hirakawacho
Chiyoda-ku, Tokyo 102
Japan

Thomas Nelson Australia
102 Dodds Street
South Melbourne, 3205
Victoria, Australia

International Thomson Publishing Asia
60 Albert Street
#15-01 Albert Complex
Singapore 189969

Nelson Canada
1120 Birchmount Road
Scarborough, Ontario
Canada M1K 5G4

International Thomson Publishing GmbH
Königswinterer Strasse 418
53227 Bonn
Germany

Printed in Canada

5 4 3 2 1

ISBN 0-534-35924-8

To the memory of Lilly C. Appel,
a constant in a world of variation,
whose love of mathematics and
teaching inspired many.

Table of Contents

Preface

The Changing Environment of Statistics Education

In the past decade or so, educators have come to reconsider the best approach to teaching and learning in applied statistics courses. With the widespread availability of personal computers, advances in statistical software, and the near-universal application of quantitative methods in many professions, courses now emphasize statistical reasoning more than computational skill development. Questions of how have given way to more challenging questions of why, when, and what?

Simultaneously, undergraduates are increasingly comfortable with software, expecting to use computers in their work. Colleges are seeking ways to integrate information technology efficiently into coursework. The introductory statistics course is an ideal place to augment traditional out-of-class assignments with structured computer exercises.

The goal of this book is to supplement an introductory undergraduate statistics course with a comprehensive set of self-paced exercises. Students can work independently, learning the software skills outside of class, while coming to understand the underlying statistical concepts and techniques. Instructors can teach statistics and statistical reasoning, rather than algebra or software.

The Approach of this Book

The book reflects the changes described above in several ways. First, and most obviously, it provides some training in the use of a powerful software package to relieve students of computational drudgery. Second, each session is designed to address a statistical issue or need, rather than to feature a particular command or menu in the software. Third, nearly all of the datasets in the book are real, reflecting a variety of disciplines. Fourth, the sessions follow a traditional sequence, making the book compatible with many texts. Finally, as each session leads the student through the techniques, it also includes thought-provoking questions and challenges, engaging the student in the processes of statistical reasoning. In designing the lab exercises, I kept four ideas in mind:

- *Statistical reasoning, not computation, is the goal of the course.* This manual asks students questions throughout, balancing software instruction with reflection on the meaning of results.
- *Students arrive in the course ready to learn statistical reasoning.* They need not slog all the way through descriptive techniques before encountering the concept of inference. The exercises invite students to think about inferences from the start, and the questions grow in sophistication as students master new material.
- *Exploration of real data is preferable to artificial datasets.* With the exception of the famous Anscombe regression dataset, all of the datasets are real. Some are very old and some are quite current, and they cover a wide range of substantive areas.
- *Statistical topics, rather than software features, should drive the design of each lab session.* Each lab session features several Minitab functions selected for their relevance to the statistical concept under consideration.

This book provides a rigorous but limited introduction to the software. Minitab is rich in features and options; this book makes no attempt to "cover" the entire package. Instead, the level of coverage is commensurate with an introductory course. There may be many ways to perform a given task in Minitab; generally, I show one way. This book provides a "foot in the door." Interested students and other users can explore the software possibilities via the extensive Help system or other standard Minitab documentation.

The Datasets

As previously noted, each of the datasets provided with this book contains real data, much of it downloaded from public sites on the World Wide Web. Appendix A describes each file, its source, and provides detailed definitions of each variable.

The data files were chosen to represent a variety of interests and fields, and to illustrate specific statistical concepts or techniques. No doubt, each instructor will have some favorite datasets which can be used with these exercises. Most textbooks provide datasets as well. For some tips on converting other datasets for use with Minitab, see Appendix B.

Note on Software Versions

The examples in this manual are based on Minitab Version 12, running under Windows 95 or Windows NT. Users of earlier Windows versions or the Student Version will notice only minor differences with the figures and instructions in this book, and in a few instances, will need to take an alternate approach. Appendix D addresses instances where Version 12 deviates substantially from earlier releases.

Version 12 does introduce some new features which are valuable in the introductory course, and therefore figure in some of the sessions. In particular, one- and two-sample tests for population proportions are now available and appear in Sessions 10 through 12. Users of earlier versions will need to make adjustments in parts of these sessions.

Acknowledgments

This project enjoyed the support of the Stonehill Undergraduate Research Experience (SURE) program during the summer of 1996, as well as the institutional support of Stonehill College. Nan Mulford and Kathy Conroy of the College's Office of Academic Development were a source of competent, congenial, and thoughtful assistance.

As the SURE Scholar on the project, Debra Elliott was a full partner in the conception of the lab assignments and in the selection and preparation of the datasets. Deb's skill, care and diligence were of enormous value in the project, as were her independence and insights into student concerns and interests.

Many colleagues and students have contributed to the creation of this book. Professors Roger Denome and James Kenneally of Stonehill

College steered me to interesting datasets in their fields of study, and Jim constantly encouraged my work. Professor Annie Puciloski suggested many improvements to the manuscript. Among my students, Jaime Annino and Bevin Ronayne skillfully field-tested many of the sessions, and Nicole Lecuyer patiently tended to some particularly tedious chores, and was a frank, resourceful, and supportive reviewer. Professors Stephen Nissenbaum of the University of Massachusetts and Roger Johnson of South Dakota School of Mines and Technology were gracious in offering advice and assistance. Thanks also to the many individuals and organizations granting permission to use the data for these exercises; they are all identified in Appendix A.

Minitab's Author Assistance Program, through the good works of Jeff Hartzell and David McClelland, certainly lived up to its name.

At Duxbury Press and Brooks/Cole, May Clark ably has helped with many questions and Cynthia Mazow has been a thoroughly dependable and affable guide and partner throughout the project. It has been my good fortune to work with them both. Thanks also to the following reviewers whose constructive suggestions have improved the quality of this book: Gregory Davis, University of Wisconsin, Green Bay; Robert Fountain, Portland State University; Jeffrey Jarrett, University of Rhode Island; Dennis Kimzey, Rogue community College; Roxy Peck, California Polytechnic State University; and Bruce Trumbo, California State University, Hayward. The remaining flaws are all mine.

Finally, my thanks go to the home team: Donna, Sam and Ben. Help comes in many forms, but none more valued or deeply appreciated than the kind they have provided throughout this experience.

To the Student

This book has two goals: to help you understand the concepts and techniques of statistical analysis, and to teach you how to use one particular tool—Minitab—to perform such analysis. It can supplement but not replace your principal textbook or your classroom time. To get the maximum benefit from the book, you should take your time and work carefully. Read through a session before you sit down at the computer. Each session should require no more than about 30 minutes of computer time; there's little need to rush through them.

You'll often see questions interspersed through the computer instructions. These are intended to shift your focus from 'getting answers' to thinking about what the answers mean, whether they make sense, whether they surprise or puzzle you, or how they relate to what you have been doing in class. Attend to these questions, even when you aren't sure of their purpose.

As noted earlier, Minitab is a large and very powerful software package, with many capabilities. Many of the features of the program are beyond the scope of an introductory course, and do not figure in these exercises. However, if you are curious or adventurous, you should explore the menus and Help system. You may find a quicker, more intuitive, or more interesting way to approach a problem.

Typographical Conventions

Throughout this manual, certain symbols and typefaces are used consistently. They are as follows:

Menu ➢ Sub-menu ➢ Command The "mouse" indicates an action you take at the computer, using the mouse or keyboard. The bold type lists menu selections for you to make.

`Variable names and items you should type appear in this typeface.`

File names (e.g., Colleges) appear in this typeface.

> A box like this contains information about something which may work differently on your computer system.

Bold, italics in the text indicate a question which you should answer as you write up your experiences in the lab.

Each lab ends with a section called "***Moving On...***." You should also respond to the numbered questions in that section, as assigned by your instructor.

Session 1

A First Look at Minitab

Objectives

In this session, you will learn to do the following:

- Launch and exit Minitab
- Enter quantitative and qualitative data in a worksheet
- Create and print a graph
- Get Help
- Save your work to a diskette

Launching Minitab

Before starting this session, you should know how to run a program within the Windows 95 or Windows NT operating system. All of the instructions in this manual presume basic familiarity with the Windows environment.

> Check with your instructor for specific instructions about running Windows 95/NT on your system. Your instructor will also tell you where to find Minitab.

Click and hold the left mouse button on the **Start** button at the lower left of your screen, and drag the cursor to select **Programs**. Locate and choose **Minitab 12 for Windows**. Drag over to **Minitab**, and release the mouse button to launch the program.

Because Minitab is a large program, you will have to wait a few moments before the program is ready for use. On the next page is an image of the screen you will see when Minitab is ready. There is a menu bar across the top of the screen and two open windows. The upper window is the *Session Window,* which will contain the results of all commands you issue to Minitab. The lower window is the *Data Window* which is used to display the data which you will analyze using the program. Later, you will see that it is possible to have several open data windows, showing multiple worksheets. For now, we'll work with just one. Each window has a unique purpose, and each can be saved separately to disk. It's important at the outset to have a sense of what each window is about.

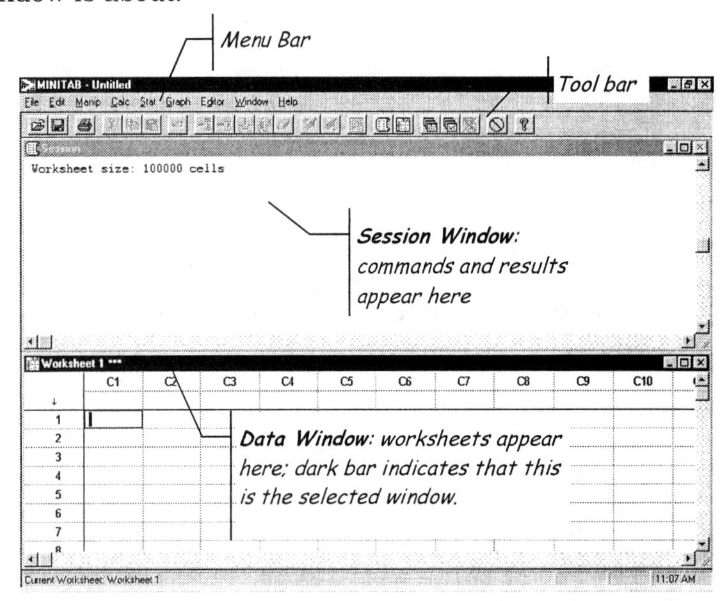

At any point in your session, only one of the Windows is *selected,* meaning that mouse actions and keystrokes will affect that window alone. To select a window, place the cursor anywhere in the window, and click the left button once. When you start Minitab, the Data Window is initially selected.

Since Minitab operates upon data, we generally start by placing data into a worksheet, either from the keyboard of from a stored disk file. The Data Window looks much like a spreadsheet. Cells may contain numbers or text, but unlike a spreadsheet, they never contain formulas. Columns are numbered C1, C2, and so on. Except for the first row, which is reserved for variable names, rows are numbered consecutively. Each variable in your dataset will occupy one column of the Minitab worksheet, and each row represents one observation. For example, if you have a sample of fifty observations on two variables, your worksheet would contain two columns and fifty rows.

There are also two other windows which are minimized when Minitab starts, so you do not see them: the Info and History Windows. The *Info Window* contains summary information about current active worksheets. Specifically, it lists the columns containing data, the names assigned to the variables, the number of observations in each column, and the number of missing values in each column.

The *History Window* records all commands issued by you during a session. As you select commands from the menus, History provides a complete transcript of all of your actions. By using the cut and paste features under the Edit menu, you can re-issue or slightly modify commands, saving time.

You can also preserve a record of all commands right within the Session Window itself. By default, Minitab only shows the *results* of computations or commands in the Session Window. To be able to enter commands, or edit results, you should do the following whenever you run Minitab:

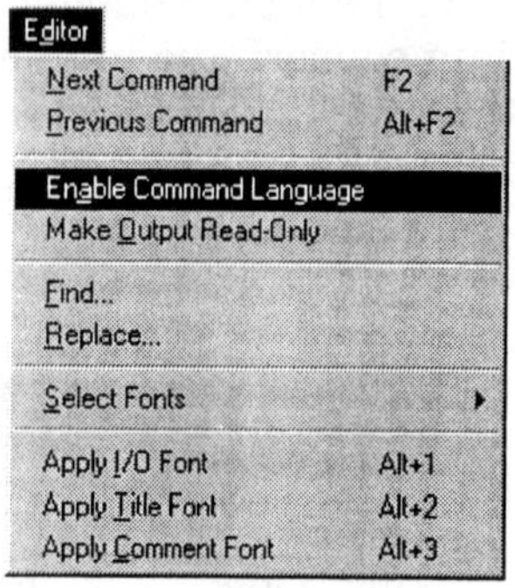

- Click anywhere in the Session Window to make it the active (or selected) window.

- On the **Editor** menu, select **Enable Command Language**. This will ensure that future menu selections appear within the Session Window.

You should get into the habit of issuing this command each time you start a lab session. In addition you should type your name into the Session Window so that you can identify your printed output later.

As you use Minitab, other windows may open as you create graphs or issue various commands. Any window can be enlarged, shrunk, or closed altogether by clicking on the small icon boxes in the upper right corner of the window, or by clicking and dragging the lower right corner of the window.

The menu bar across the top of the screen identifies broad categories of Minitab's features. There are two ways to issue commands in Minitab: choosing commands from the menu or icon bars, or typing them directly into the Session Window using Session commands. This book always refers you to the menus. As you make menu choices, Minitab actually "writes" a command into the Session and History Windows, so even with menus you are actually using Session Commands.[1] You can do no harm by clicking on a menu and reading the choices available, and you should expect to spend some time exploring your choices in this way. As your skills develop and you become more familiar with the software, you may want to use icons or Session Commands.

Entering Data into a Worksheet

For most of the sessions in this manual, you will start by accessing data already stored on a disk. For small datasets, though, it will often make sense simply to type in the data yourself. For this session, you will transfer the data displayed below into the empty worksheet in the Data Window.

In this first example, our goal is simple: to create a small worksheet, and then use Minitab to construct two graphs from the data. This is typical of the tasks you will perform throughout the book.

The data concern the total number of Acquired Immune Deficiency Syndrome (AIDS) cases reported to the World Health Organization (WHO) through 1994. This is an example of *cross-sectional* data: for each continental region, at one point in time, we record the total number of AIDS cases, as well as the number of countries filing a report with WHO.

Region	# of Countries	Total Cases Reported
Africa	54	347713
Americas	45	526682
Asia	41	17057
Europe	38	127886
Oceania	14	5735

The first row in a worksheet, above Row 1, is reserved for variable names. Although Minitab does not require us to use variable names, we almost always will do so for our convenience. In the absence of a name, each variable is simply identified by the column number: C1, C2, etc.

---

[1] It visibly writes the command in the Session Window only if you have enabled the command language, as described earlier.

- Move the cursor into the Data Window, and position it in the cell just below the heading **C1**. Type the variable name **Region** and press the **Enter** key. The cursor will move down into Row 1. Type **Africa** and then complete the column, using the data from the table on the previous page.

Note that the **C1** at the top of the column has changed to **C1-T**, indicating that this column refer to a Text (i.e., qualitative) variable.

- Title the second variable **Country**, and enter the data into Rows 1 through 5 within the second column.

- Do the same for the third variable, calling it **TotCases**.

When you are done, the worksheet in the Data Window should look like this:

Worksheet 1 ***

	C1-T	C2	C3	C4
↓	Region	Country	TotCases	
1	Africa	54	347713	
2	Americas	45	526682	
3	Asia	41	17057	
4	Europe	38	127886	
5	Oceania	14	5735	
6				
7				

Saving a Worksheet

It is wise to save all of your work in a disk file. Minitab distinguishes among several objects which one might want to save. As time goes by, you might want to save a Session or History Window, a worksheet, or a collection of related items, which Minitab refers to as a *Project*. For further discussion and description of these options, consult Appendix B.

At this point, we've created a worksheet and ought to save it on a diskette. Let's call the worksheet AIDS1.

> Check with your instructor to see if you can save the worksheet on a hard drive or network drive in your system.

- On the **File** menu, choose **Save Current Worksheet As....** In the **Save in** box, select **3 ½ Floppy (A:)**. Then, next to **File Name**, type **AIDS1**.

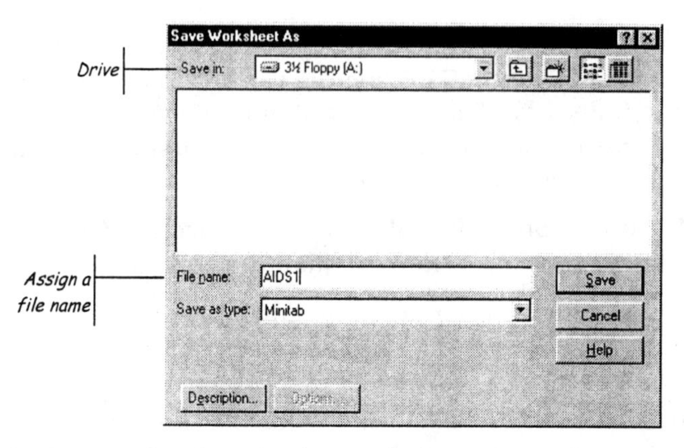

This *just* saves the data in the worksheet. Your Session and other windows have *not* been saved. That's a skill covered in a couple of pages.

Creating a Chart

WHO has summarized AIDS reports into five continental regions, combining North and South America, and combining Australia and Antarctica into a region called "Oceania." We'll use a bar graph to display the number of countries reporting from each of the regions.

Click on **Graph** in the menu bar, and choose **Chart**.

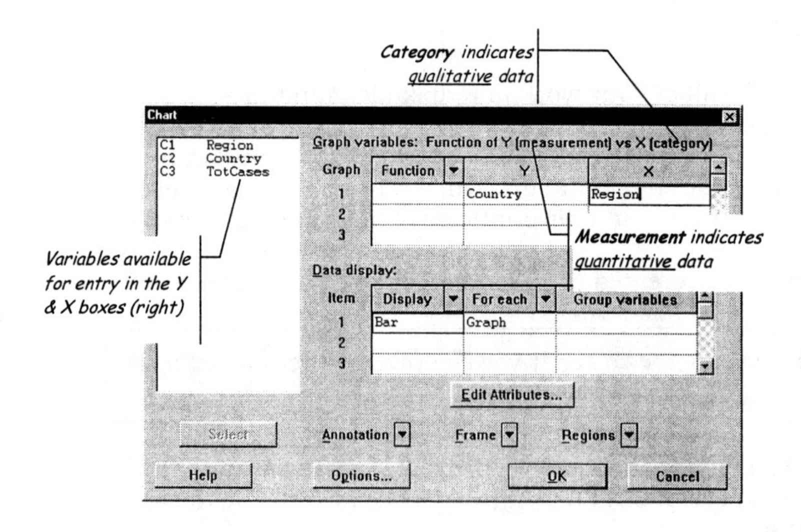

> From now on in this manual, a menu selection will be abbreviated with the name of the menu and the submenu or command. In this case, it would be **Graph ➤ Chart**.

- Move the cursor to the cell under **Y**, in the row for **Graph 1** and click once. You'll see an insertion point appear in the box. Double-click on `C2 Country` in the list at the left. Note that `C1` is missing from the list.

- Move to the next cell under **X**. Select `C1 Region` for the categorical variable.

- Now move the cursor onto the downward arrow next to the word **Annotation** and click. Select **Title** from the list. Click in the top rectangle, and type the words `Countries by Region` followed by your name in parentheses. Click **OK** in the Annotation dialog, and then **OK** again in the Chart dialog.

You will see a new window appear, containing a bar chart (see below). This chart displays the number of countries filing WHO reports, by continental region. ***What, if anything, strikes you as noteworthy about the chart? What does it tell you about the prevalence of AIDS in different parts of the world?***

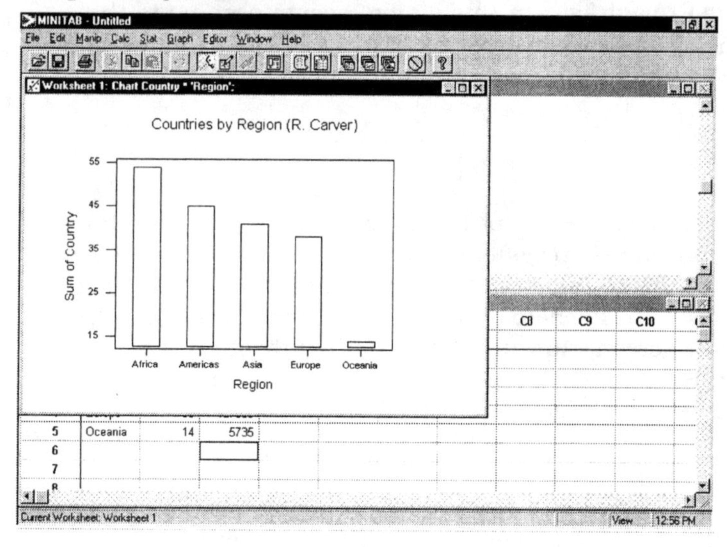

In fact this chart tells us nothing relevant to the second question. To answer that question, we should chart the number of cases reported in each region. *Before doing so, take a moment to think about what you expect this chart to look like, having seen the other one. How do you expect the two charts to compare?*

Graph ➢ Chart Notice that the dialog box still contains your last selections. This time, use `TotCases` instead of `Country` for **Y**.

Give this graph a new title, reflecting its content. Include your name in the title.

What strikes you most about this new chart? How does it compare to the first one? Why are they different?

Saving a Project

At this point, we have two open graphs, a Session Window with some commands, and a Data Window with a worksheet. We could save each window in a separate disk file before ending this session. Alternatively, we can save them all as a Minitab *Project File.* As noted earlier, a project is a related collection of worksheets, Session, History, Info and Graph windows. To save a project, do the following:

File ➢ Save Project As... In this dialog, assign a name to the file (`Session 1`). This new file will save all of the windows you see, as well as the History and Info Windows.

Getting Help

You may have noticed the "Help" button in a dialog box. Minitab features an extensive on-line Help system. If you aren't sure what a term in the dialog means, or how to interpret the results of a command, click on Help. You can also search for help on a variety of topics via the Help menu at the top of your screen. As you work your way through the sessions in this book, Help may often be valuable. Spend some time experimenting with it before you genuinely need it.

Printing in Minitab

Now that you have created some graphs, let's print one of them.

To print a graph, first select the graph of interest by clicking somewhere on the graph window.

Check with your instructor concerning any special considerations in selecting a printer or issuing a print command. Every system works differently in this matter.

File ➤ Print Graph... This command will print the selected window. Click **OK**.

In general, to print all or part of a window, you must make the window active by clicking on it, and then find the appropriate print command in the **File** menu.

Quitting Minitab

When you have completed your work, it is important to exit the program properly. Virtually all Windows programs follow the same method of quitting.

File ➤ Exit You will generally see a message asking if you wish to save changes to this project. Since we saved everything earlier, click **No**.

That's all there is to it. Later sessions will explain menus and commands in greater detail. This session is intended as a first look; you will return to these commands and others at a later time.

Session 2

Tables and Graphs for One Variable

Objectives

In this session, you will learn how to do the following:

- Retrieve data stored in a Minitab worksheet
- Create a Dotplot
- Explore data with a Stem-and-Leaf display
- Create and customize a histogram
- Create a frequency distribution
- Print output from the Session Window
- Create a bar chart

Opening a Worksheet

In the previous session, you created a Minitab worksheet by entering data into the Data Window. In this lab, you'll use several worksheets that are available on your diskette. This session begins with some detailed data about AIDS cases around the world. Our goal is to get a sense of how prevalent the disease is.

Choose **File ➤ Open Worksheet**. A dialog box like the one shown on the next page will open. In the **Look in**: box, select **3 ½ Floppy (A:)** and you will see a list of available worksheet files. Select the one named **Aids**. (This file name may appear as Aids.MTW on your screen, but it's the same file.)

NOTE: The location of Minitab files might be different on your computer system. If you have a problem, check with your instructor.

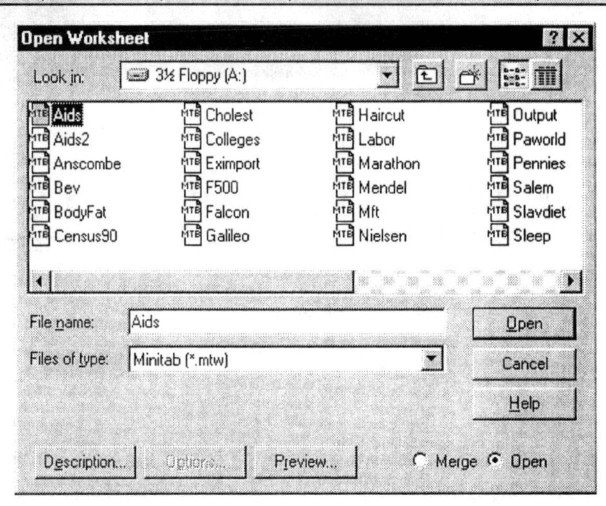

Click **Open**, and you may see the message shown below, alerting you that your current project will be augmented with this file. If you do, click **OK**, and your Data Window will show the data from the AIDS file. Using the *scroll bars* at the bottom and right side of the screen, move around the worksheet, just to look at the data. Consult Appendix A for a full description of the data files.

A Dotplot

A Dotplot is a good tool for a first look at the shape and spread of quantitative data.

Choose **Graph ➤ Dotplot**. In the dialog box, move your cursor to the list of variables on the left of the box, and click on

Case94, which represents the number of new cases reported in 1994. Then click on **Select**. You will see the word **Case94** appear in the **Variables** box. Add a title, as shown in the dialog, and click **OK**.

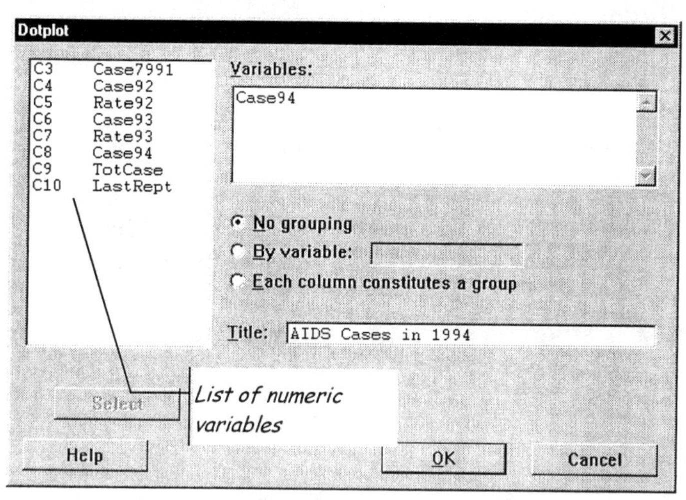

After clicking **OK**, another window (shown below) will appear. In this plot, the horizontal axis represents a number of AIDS cases, and the vertical axis represents the number of countries reporting that many cases. Each dot in the graph represents 1 or 2 countries.

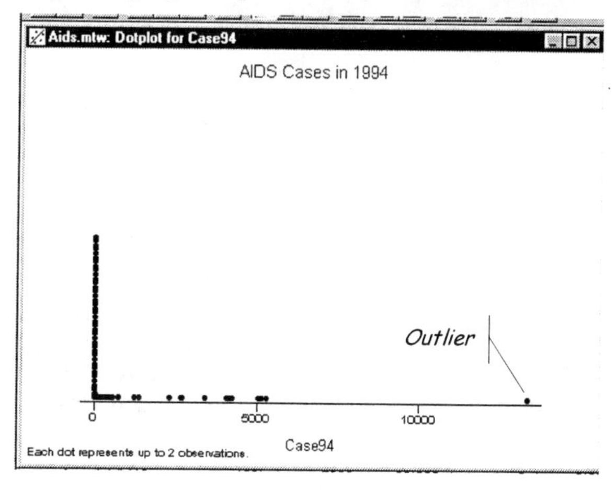

Notice that the vast majority of countries appear on the left end of the plot. ***How would you describe the shape of this distribution?***

Also notice the one dot at the right end. ***What does it mean? What country might it represent? See if you can locate that one outlier in the Data Window.***[1]

Exploring the Data with Stem-and-Leaf

Another simple tool for exploring data is a Stem-and-Leaf plot.

Stat ➤ EDA ➤ Stem-and-Leaf Select `Case94`, then click **OK**; the Stem-and-Leaf diagram will be in the Session Window. You may want to *maximize* the session window for a better look.

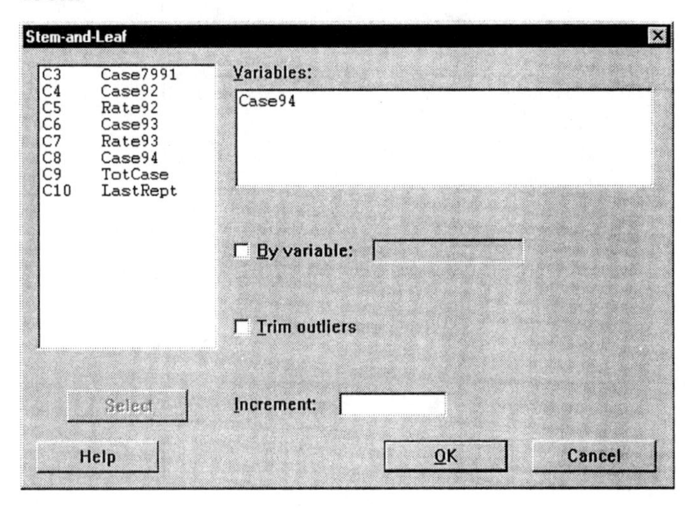

In this output, we see first that there are 208 observations, and that each stem entry represents 1000 AIDS cases. In the display, there are three columns of information, representing cumulative frequency, stems, and leaves. The first column shows cumulative frequency up to the median, and then shows declining cumulative frequency. In this example, the median—denoted by parentheses around the value—occurs in the very first class. Below that, cumulative frequencies "count down."

---

[1] Suggestion for ambitious students: On Minitab's Help menu, search for Help on "Brush" to learn how to "point" to an outlier and locate its value in the data window.

Character Stem-and-Leaf Display

```
Stem-and-leaf of Case94     N  = 208
Leaf Unit = 100


 (195)   0 00000000000000000000000000000000000000000000000000000000000000000000000+
   13    1 23
   11    2 266
    8    3 3
    7    4 012
    4    5 013
    1    6
    1    7
    1    8
    1    9
    1   10
    1   11
    1   12
    1   13 3
```

This diagram shows that 195 countries had fewer than 1000 cases in 1994. Thirteen countries had 1000 or more, 11 had 2000 or more, and so on. If you look at the second row, with a stem value of 1 (indicating 1000's), what do the 2 and 3 indicate? One country had about 1200 cases, and another about 1300. In the next row, one country had about 2200, and two countries had about 2600. Skip down to the stem value of 6; there are no leaves, meaning that there were no countries in the 6000 – 6999 case range.

The Stem-and-Leaf display is a tool for developing meaningful frequency distributions, and provides a crude visual display of the data.

Creating a Histogram

Graph ➤ Histogram. Select `Case94`. Click on **Annotation** and select **Title** from the drop-down list. In the dialog box, type a title for this graph (e.g., "1994 AIDS Cases Reported"). Click **OK**, to accept the title. Then choose **Footnote** under **Annotation**, and place your name in a footnote to the graph. Click **OK** and then click **OK** again, and a new window will open, containing the graph.

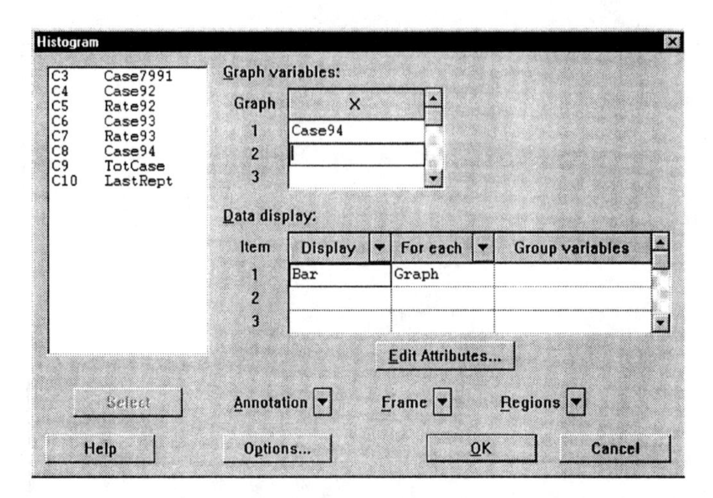

Compare this histogram to the dotplot and stem-and-leaf displays. ***What important differences, if any, do you see?***

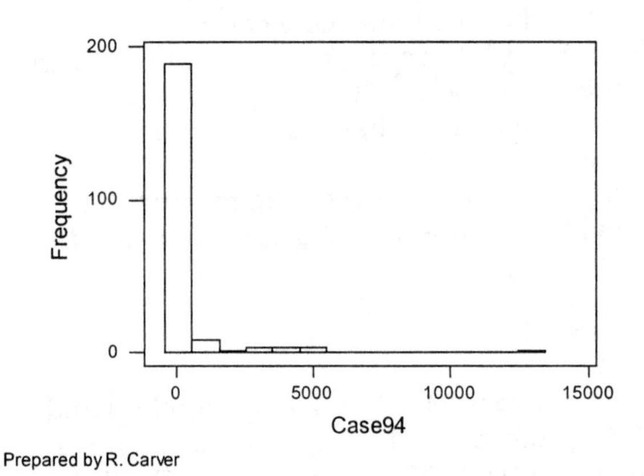

In this histogram, Minitab determined the number of bars, which affects the apparent shape of the distribution. We can control the number of bars as follows:

Edit ➤ Edit Last Dialog This immediately brings you back to the most recent dialog, and you can make the necessary changes. In this case, you need to click on the **Options...** button. This brings up a new dialog box (shown below).

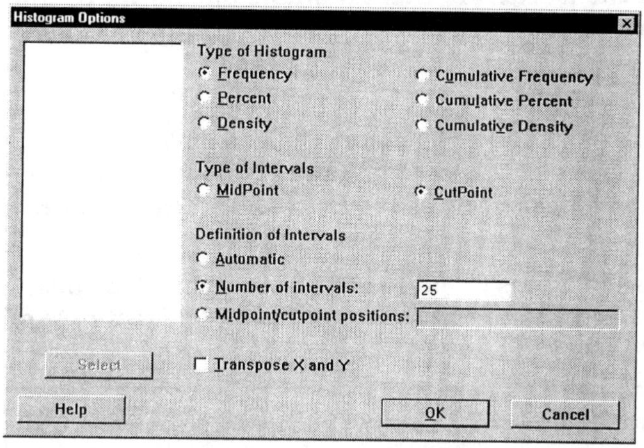

Histogram Options

Type of Histogram
Frequency
Percent
Density
Cumulative Frequency
Cumulative Percent
Cumulative Density

Type of Intervals
MidPoint
CutPoint

Definition of Intervals
Automatic
Number of intervals: 25
Midpoint/cutpoint positions:

Select
Transpose X and Y

Help
OK
Cancel

Change the **Type of Intervals** from **Midpoint** to **Cutpoint** as shown in the dialog above. This resets the relationship of the bars to the horizontal axis, so that the endpoints of class intervals are labeled.

Click in the **Number of Intervals** box, and type 25. Then click **OK** to exit the **Options** dialog, and **OK** again to create the revised histogram.

How does this compare to your first one? Which graph better summarizes the dataset?

We would ordinarily expect more populous countries to have more AIDS cases than smaller countries. As such, it might make more sense to think in terms of the *proportion* of the population suffering from the disease in each country. In our dataset, we don't have a "grand total" rate of incidence, but we do have annual rates for several years. Consider the variable `Rate93`. This is the number of AIDS cases reported in 1993 per 100,000 people. In other words, if a country with a population of 10 million people reported 15 new cases of AIDS, they would have a rate of 1.5 per 100,000.

Construct a histogram for the variable `Rate93`.

Looking at the histogram, how would you describe the shape of this distribution? Approximately which rates of incidence seem most common around the world? Is there a single outlier as in the earlier analysis? In which country or countries is the disease most prevalent?

Now, let's see a graph of the *cumulative* frequency distribution. Find your way back to the **Histogram Options** dialog. At the top of the options dialog box, find **Type of Histogram**; press the options button for **Cumulative Frequency**. Click **OK**, then **OK** again.

Compare the results of this graph to the prior graph. ***What does this one show? About how many countries had a 1993 AIDS rate of less than 50 cases per 100,000 population?***

> NOTE: You now have several graph windows open, which consumes a lot of memory. To avoid memory problems, close one or more of them. You can do this one graph at a time by clicking on the Control Menu box in the upper left corner of the graph window, or by selecting **Window ➤ Close all graphs** or **Window ➤ Manage Graphs**.

Frequency Distributions with Tally

Let's switch to another worksheet and a set of questions concerning *qualitative data*. First, close the AIDS file:

Select the Data Window by clicking anywhere in it.

File ➤ Close Worksheet You'll see a message alerting you that this worksheet will now be removed from your project. Click **OK**.

File ➤ Open Worksheet. Choose the worksheet called **Census90.**

This file contains a random sample of 982 Massachusetts citizens, with their responses to selected questions on the 1990 United States Decennial Census. One of the questions on the Census form asked how they commute to work. In our dataset, the relevant variable is called

"Trans" (means of transportation) and occupies C13. This is a *categorical*, or *nominal*, variable. The Bureau of Census has assigned the following code numbers to represent the various categories:

Value	Meaning
0	n/a, not a worker or in the labor force
1	Car, Truck, or Van
2	Bus or trolley bus
3	Streetcar or trolley car
4	Subway or elevated
5	Railroad
6	Ferryboat
7	Taxicab
8	Motorcycle
9	Bicycle
10	Walked
11	Worked at Home
12	Other Method

To see how many people in the sample used each of these means of transportation, we can have Minitab generate a simple tally.

Stat ➤ Tables ➤ Tally. In the Tally dialog box, select variable **`Trans`** and click **OK**. The tally results appear in the session window.

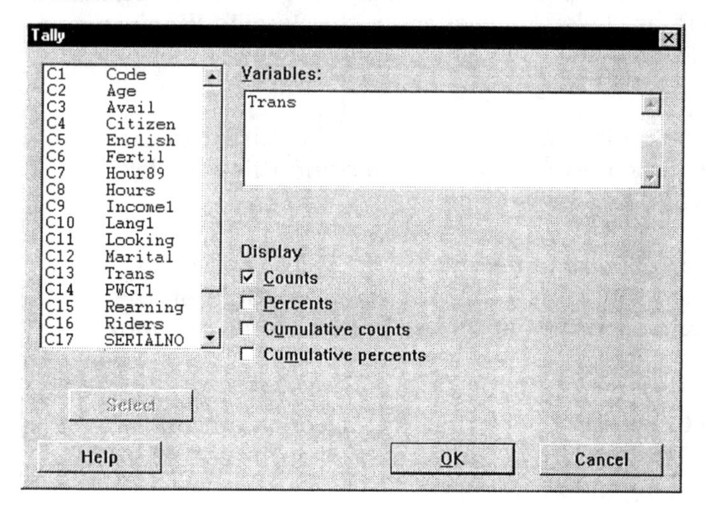

In the session window, you should now see this:

Summary Statistics for Discrete Variables

Trans	Count
0	556
1	354
2	10
4	1
5	2
7	2
9	1
10	28
11	22
12	6
N=	982

Note: *the "Trans" column shows values of the variable, and the "Count" column shows frequencies.*

Among people who work, which means of transportation is the most common? The least common? Be careful: the most common response was 0, which represents someone who does not work at all.

Printing Session Output

Sometimes you will want to print all or part of a session window. Before printing your session, be sure you have made session output editable (see Session 1), and typed your name into the session. You can print the entire session by clicking anywhere in the Session Window, and then choosing **File ➤ Print Window**. To print *part* of a Session Window, do this:

- In the Session Window, move the "I-beam" cursor to the beginning of the portion you want to print. Click the left mouse button.

- Using the scroll bars (if necessary), move the cursor to the end of the portion you want to print. Then press Shift on the keyboard and click the left mouse button. You'll see your selection highlighted.

- **File ➤ Print Window**. Notice that the **Selection** button is already marked, meaning that you'll print a selection of the entire Session Window. Click **OK**.

Another Bar Chart

To graph this tally, we should make a bar chart.

- **Graph ➤ Chart**. In the dialog box (see below), you see that Minitab requests three pieces of information for Graph 1: a function, a Y variable, and an X variable. To represent one qualitative variable (X), we only need to fill in the **X** box. In this case, we want to know how many people gave each answer for the variable called `Trans`.

- Move your cursor to the X variable column, and click. Then select the variable `C13 Trans`. Give the graph a title (**Annotation**), and click **OK**. Click **OK** again to make the graph.

Compare the chart and the tally. They should contain the same information. Do they?

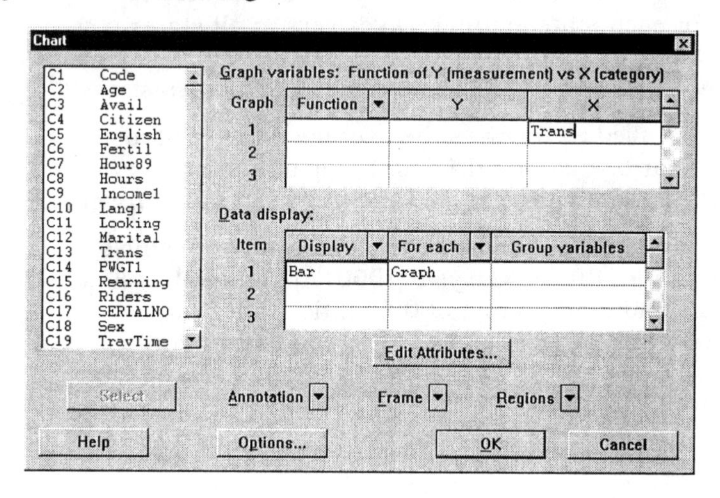

Moving On...

Using the skills you have practiced in this session, now answer the following questions. In each case, provide an appropriate graph or table to justify your answer, and explain how you draw your conclusion.

1. Note that the `Trans` variable includes the responses of people who don't work outside the home. Among those who do have to commute, what proportion use some type of public transportation (codes 2 through 5)?

For the questions below, you will need to use the files **States**, **Marathon**, and **Nielsen** (see Appendix A for detailed file descriptions). You may be able to use several approaches or commands to answer the question; think about which approach seems best to you.

States

2. The variable named `BAC` refers to the legal blood alcohol threshold for driving while intoxicated. All states set the threshold at either .08 or .10. About what percentage of states use the .08 standard?
3. The variable called `INS94` is the average auto insurance premium for each state in 1994. Do drivers in all states pay about the same amount for insurance? What seems to be a typical amount? How much variation is there across states?
4. The variable called `Mileage` is the average number of miles driven per year by a state's drivers. Using a dotplot, can you find two states where drivers lead the nation in miles driven?
5. The variable called `AccFat` is the number of motor vehicle accident fatalities during the year 1990. Approximately how many states had more than 1,500 fatalities that year? {Presumably, this variable is related to population and mileage; in Session 3, we'll see how to take that relationship into account.}

Marathon

This file contains the finish times for the wheelchair racers in the 1996 Boston Marathon.

6. The variable `Country` is a three-letter abbreviation for the home country of the racer. Not surprisingly, most of the racers were from the USA. What country had the second highest number of racers? (Hint: Make a chart.)

7. Approximately what percentage of wheelchair racers completed the 26-mile course in less than 2 hours, 10 minutes (130 minutes)?

8. How would you characterize the shape of the histogram of the variable `Minutes`? (Experiment with different numbers of intervals in this graph.)

Nielsen

This file contains the Nielsen ratings for the twenty most-heavily-watched television programs for the week ending September 14, 1997.

9. Which of the networks reported had the most programs in the top 20? Which had the fewest?

10. Approximately what percentage of the programs enjoyed ratings in excess of 11.5?

Session 3

Tables and Graphs for Two Variables

Objectives

In this session, you will learn how to do the following:

- Represent two variables in different ways in a worksheet
- Cross-tabulate two variables
- Create a Bar Chart comparing two variables
- Create a histogram for two variables
- Create an X-Y scatter plot for two quantitative variables

Cross-Tabulating Data

In the first example, we will consider a case of two qualitative variables. The example involves the Census data which you saw in the last session, and in particular deals with the question: "Do men and women use the same methods to get to work?" Since Sex and Trans (means of transportation) are both categorical data, our first approach will be a *joint frequency table* also known as a *cross-tabulation.*

Open the Census file by selecting **File ➤ Open Worksheet,** and choosing **Census90**.

Stat ➤ Tables ➤ Cross Tabulation You will see the dialog box below. Just select the variables `Trans` and `Sex`, and click **OK**. You'll find the cross-tabulation in the Session Window. Recall that Car, Truck, or Van is coded with a 1 in the `Trans` variable, and that 0 stands for male in the Sex variable. ***Who makes greater use of cars, trucks or vans?***

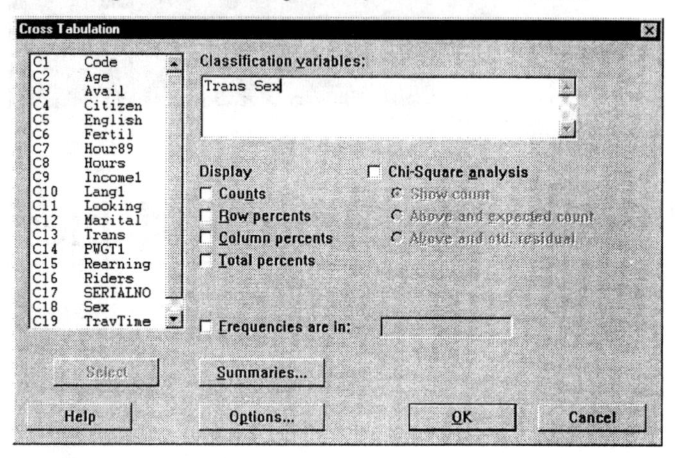

Tabulated Statistics

```
 Rows: Trans      Columns: Sex

             0         1       All

 0         258       298       556
 1         186       168       354
 2           3         7        10
 4           1         0         1
 5           2         0         2
 7           0         2         2
 9           0         1         1
 10         15        13        28
 11          6        16        22
 12          3         3         6
 All       474       508       982

  Cell Contents --
                      Count
```

The results of the cross-tabulation command may appear confusing at first, but they are not terribly mysterious. As indicated in the first two lines of output, the rows of the table represent the various means of transportation (refer again to Appendix A for the codes). The columns labeled 0 and 1 refer to males and females, respectively. Thus, for instance, 168 women commuted in a car, truck, or van.

Simply looking at the frequencies could be misleading, since the sample does not have equal numbers of men and women. It might be more helpful to compare the percentage of men commuting in this way to the percentage of women doing so. Even percentages can be misleading if the samples are small. Here, fortunately, we have a large sample. Later we'll learn to evaluate sample information more critically with an eye toward sample size.

The cross-tabulation function can easily convert the frequencies to relative frequencies. We could return to the Cross-Tabulation dialog following the same menus as before, or take a slightly different path.

Editing Your Most Recent Dialog

- **Edit ➤ Edit Last Dialog** This command will always return you to the most recent dialog box. To answer the question posed above, we want the values in each cell to reflect frequencies relative to the number of women and men, so we want to divide each by the total of each respective column. To do so, check **Column Percents**, then click **OK**. ***Based on this table, would you say that men or women are more likely to commute by car, truck, or van?***

- Now try asking for **Row Percents** instead of column percents. {**Edit ➤ Edit Last Dialog** } ***What do these numbers represent?***

More on Bar Charts

We can also use a bar chart to analyze the relationship between two variables. Let's look at the relationship between two *qualitative* variables in the Student survey: gender and seat-belt usage. One of the items (`BELT`) refers to how frequently students wear seatbelts when driving. The students' four possible responses to this were coded as follows: N = Never, S = Sometimes, U = Usually, and A = Always. What do you think the students said? Did men and women respond similarly? We will create a bar chart which will help to answer these questions.

Open the worksheet called **Student**. (you may want to close the **Census90** file before doing so, to free up some memory space).

Graph ➤ Chart The chart command was used in both prior labs. In this dialog we must specify a grouping variable (X). Optionally, we can specify a Function and a quantitative Y variable.

Move the cursor to the **X (category)** column, and select **BELT** as the variable. If we were to click **OK** now, we would see the total number of students who gave each response. But we are interested in the comparison of men and women.

In the **Data display** area, specify that you want to display a **Bar** for each **Group** (by clicking on the arrow next to **For each**, as shown below) and then select `Gender` as the **Group Variable**.

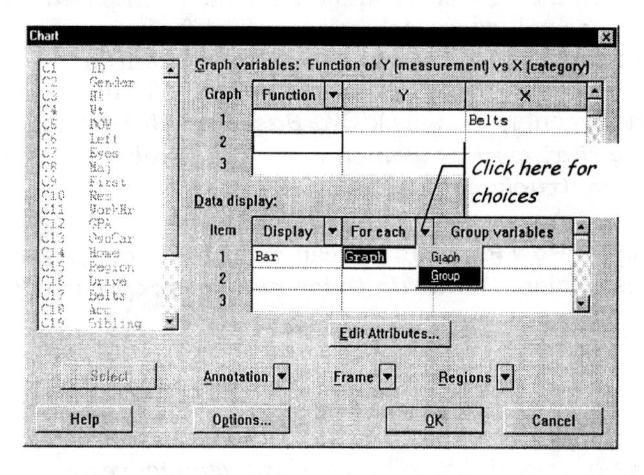

Now click on the **Options** button. In the options dialog, check **Cluster** under **Groups within X**, and select `Gender` as the Cluster variable. This will create side-by-side bars for males and females for each of the possible X-values. Click **OK**.

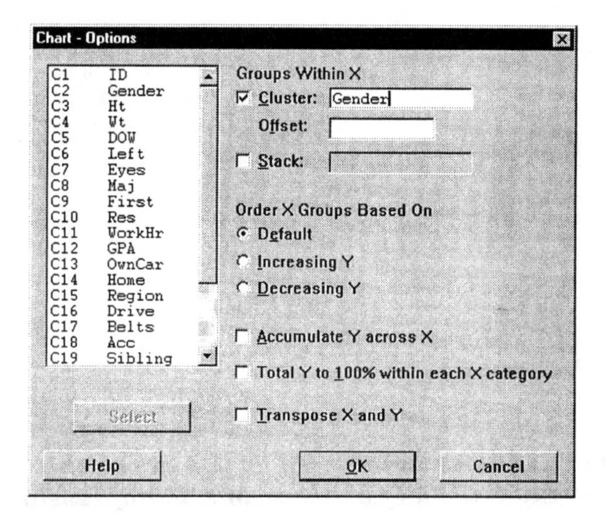

Now, click **Edit Attributes**, and make the bar (**"Back Color"**) for females **Red**, and for males, **Blue** as shown below. Click **OK**, and then **OK** for the chart.

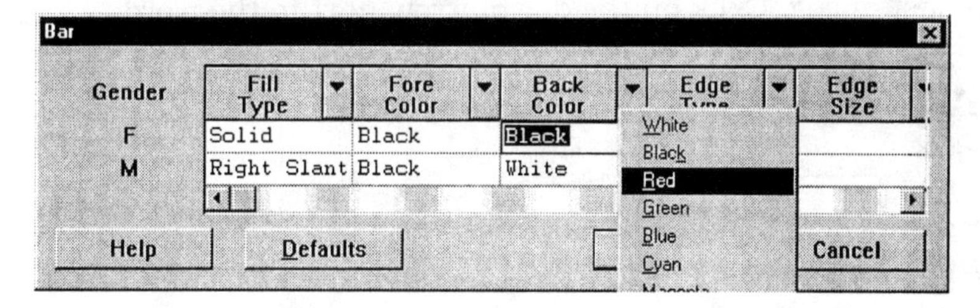

What can you say about the seatbelt-wearing habits of these students?

With the **Chart** command, we can also analyze a quantitative variable. Suppose we wanted to compare the grade-point-averages (GPA's) of the men and women in the student survey. We might consider comparing the averages of the two groups.

Graph ➢ Chart

Among the **Functions**, select **Mean**. (The mean is what you have always thought of as the "average.") Select **GPA** as the Y variable, and **Gender** as the X variable.

Return to **Options**, and uncheck **Cluster**. Click **OK**.

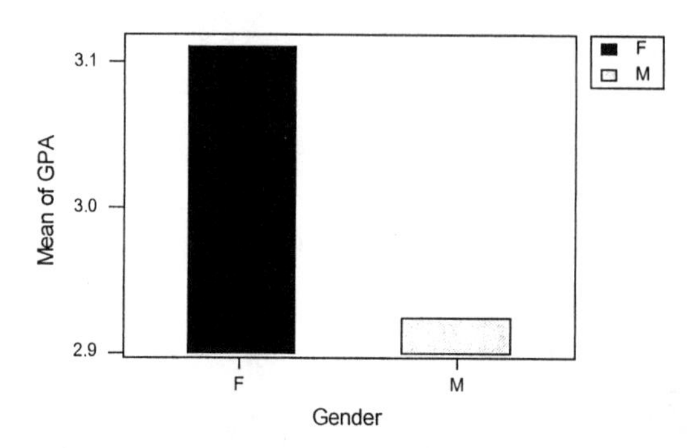

What does this graph show? What strikes you about this graph? Can you think of any reasons for the dramatic impression which this graph makes?

To create a more accurate impression, we should begin counting along the vertical axis at 0.

Edit the last dialog. Click on the down arrow next to the word **Frame**. Choose the option **Min and Max**.

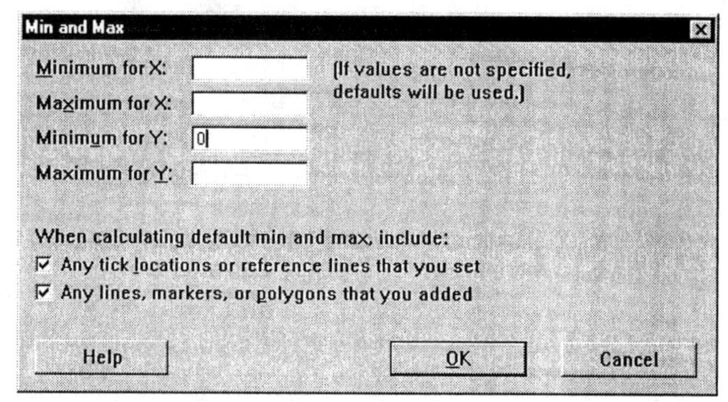

In the Min and Max dialog, type 0 in the box next marked **Minimum for Y** to have the Y-axis to start at zero. Now click **OK** for **Min and Max**, and **OK** for the **Chart**. ***How do the GPA's compare for males and females now?***

Comparing Two Distributions

This chart compares the means of two distributions. How do the entire distributions compare? You already know how to create a histogram for a quantitative variable. Let's expand our knowledge of the histogram function, using the GPA variable. We begin by looking at the distribution of GPA's across all students.

Graph ➤ Histogram... Select GPA as the variable, and click **OK**. You'll see the graph shown here. ***How would you describe the shape of this distribution?***

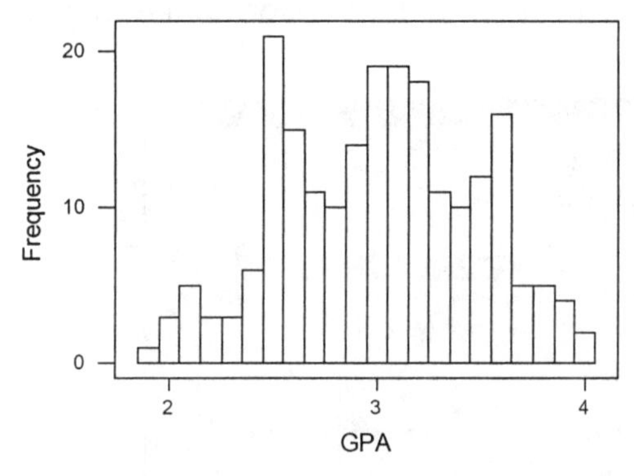

Let's see how the distribution of grades compares for male and female students. We'll create one histogram showing both distributions:

Edit ➤ Edit last dialog We need to indicate that the GPA's are to be displayed by gender, and that the graph should distinguish between the two.

To differentiate, click on the downward arrow next to the word **Display** in the Data display area of the Histogram dialog. Instead of Bar; select **Connect**.

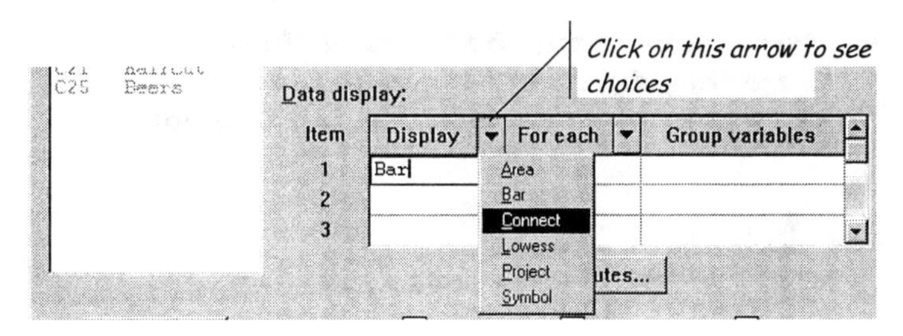

Click on the arrow next to the heading **For each**, and select **Group**.

Click in the **Group Variables** column, and select the variable `Gender` as the group variable. This will group the GPA values by gender.

Now click the **Edit Attributes** button. Click the arrow next to **Line Color**, and select **Red** for the line that will represent the women. Click on the word `Black` in the second row, and then again click the down arrow next to **Line Color**. Make this line **Blue**.

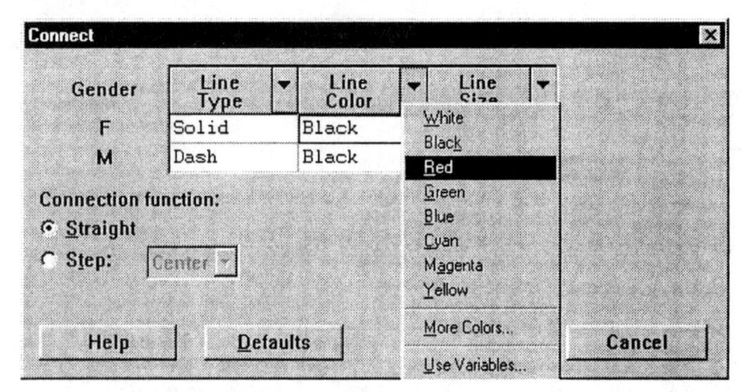

Move the cursor to the area labeled **Connection function:** Check **Step**, specifying that the lines connecting the frequency values will step up and down. Click **OK**.

Click on the **Annotation** arrow and select **Title**. Place your name on the graph for later printing. Click **OK** in the title dialog, and then **OK** to generate the histogram, a copy of which appears below.

What does this graph show about the GPA's of these students? In what ways are they different? What do they have in common? Can you think of reasons to explain the patterns you see?

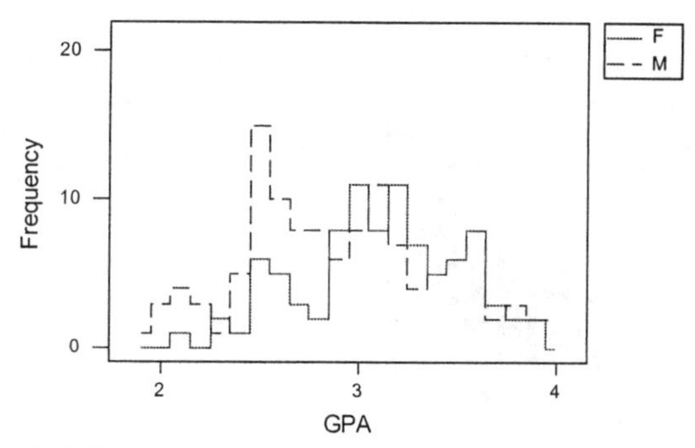

Scatterplots to Detect Relationships

The prior example involved a quantitative and a qualitative variable. Sometimes, we might suspect a connection between two quantitative variables. In the student data, for example, we might think that taller students generally weigh more than shorter ones. We can create a "scatter plot" or XY graph to investigate.

Graph ➤ Plot In the dialog box, select `WT` as the Y (vertical axis) variable, and `HT` as the X variable. Title your graph (**Title** in the **Annotation** options). Click **OK**.

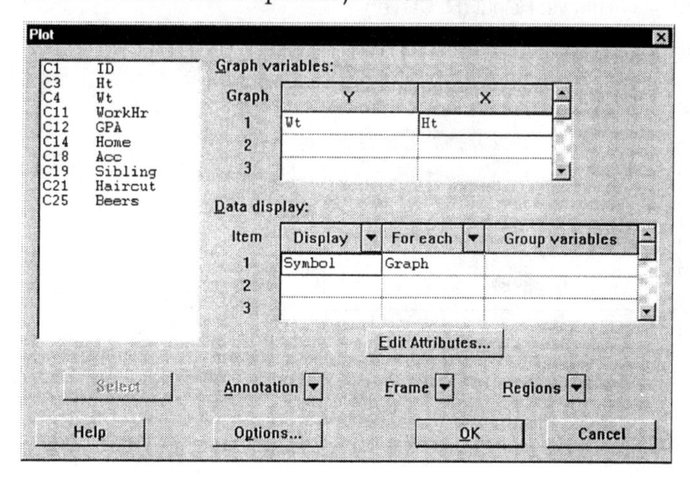

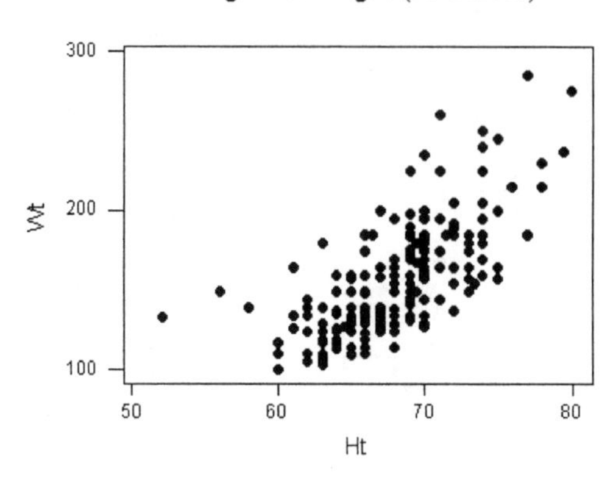

What does the graph show you? By eye, estimate the weight of a person who is 5'8" (or 68 inches) tall.

We can easily incorporate a *third* variable into this graph. Edit the last dialog, and move to the **Data** display area. This time, we'll group by `Gender`. Specify that you want to display a **Symbol** for each **Group**, with `Gender` as the **Group variables.**

Click **Edit Attributes**. Choose red for the women, and blue for the men. Click **OK** for Attributes, and **OK** in the Plot dialog.

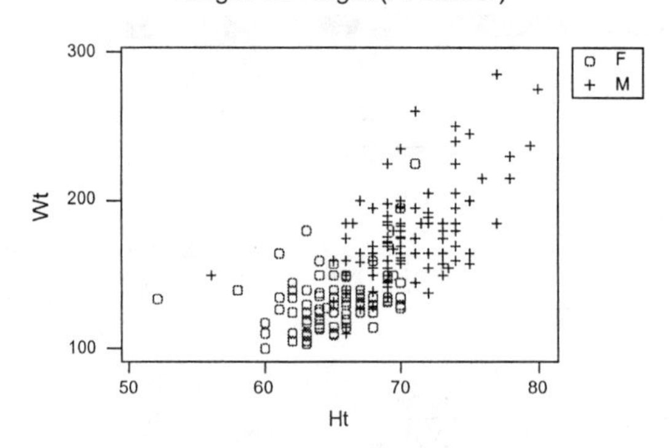

In what ways is this graph different from the first scatterplot? What additional information does it convey? Can you

make any generalizations about the heights and weights of men and women? Can you detect any outliers?

Moving On...

Create the tables and graphs described below. Refer to Appendix A for complete data descriptions. ***Be sure to title each graph, including your name. Print these completed graphs.***

Student

1. Generate a Histogram of the distribution of heights, separating men and women. Comment on the similarities and differences between the two groups.
2. Do the same for students' weights.

Bev

3. Using the Chart command, display the mean of Revenue per Employee, by SIC category. Be sure to set the minimum of the Y-axis at 0. Which of the beverage industries generates the highest average revenue by employee? (See Appendix A for a list of SIC codes).
4. Make a similar comparison of Inventory Turnover averages. Can you explain the pattern you see?
5. Do the same for Current ratios.

SlavDiet

In *Time on the Cross: The Economics of American Negro Slavery,* by Robert William Fogel and Stanley Engerman, the diets of slaves and the general population are compared.

6. Create two bar charts summing up the calories consumed by each group, by food type. Take care that the vertical axes are in comparable units. How did the diets of slaves compare to the rest of the population, according to these data?

Galileo

In the sixteenth century, Galileo conducted a series of famous experiments concerning gravity and projectiles. In one experiment, he released a ball to roll down a ramp. He then measured the total horizontal distance which the ball traveled until it came to a stop. The data from that experiment occupy the first two columns of the data file.

In a second experiment, a horizontal shelf was added to the base of the ramp, so that the ball rolled directly onto the shelf from the ramp. Galileo recorded the vertical height and horizontal travel for this apparatus as well, which are in the third and fourth column of the file.[1]

7. Construct a scatterplot for the first experiment, with release height on the X axis and horizontal distance on the Y axis. How would you describe the relationship between the two variables?

8. Do the same for the second experiment.

Aids

9. Using the Chart command, display the mean AIDS case rate in 1993, by World Health Organization region. Which region of the world had the highest incidence of AIDS cases per population in 1993?

Mendel

Gregor Mendel's early work laid the foundations for modern genetics. In one series of experiments with several generations of pea plants, his theory predicted the relative frequency of four possible combinations of color and texture of peas.

10. Construct charts of both the actual experimental (observed) results and the predicted frequencies for the peas. Comment on the similarities and differences between what Mendel's theory predicted, and what his experiments showed.

Salem

In 1692, twenty persons were executed in connection with the famous witchcraft trials in Salem, Massachusetts. At the center of the controversy was Rev. Samuel Parris, minister of the parish at Salem Village. This data file represents a list of all residents who paid taxes to

---

[1] *Sources:* Drake, Stillman. *Galileo at Work,* (Chicago: University of Chicago Press, 1978); Dickey, David A. and Arnold, J. Tim "Teaching Statistics with Data of Historic Significance," *Journal of Statistics Education,* v.3, no. 1, 1995.

the parish in 1692. In 1695, many villagers signed a petition supporting Rev. Parris.

11. Construct a crosstab of pro-Parris status and the 'accuser' variable. (Hint: compute row or column percents) Based on the crosstab, is there any indication that accusers were more or less likely than non-accusers to support Rev. Parris?
12. Construct a crosstab of pro-Parris status and the 'defend' variable. Based on the crosstab, is there any indication that defenders were more or less likely than non-defenders to support Rev. Parris?
13. Create a chart showing the mean (average) taxes paid, by Pro-Parris status. Did one group tend to pay higher taxes than the other?
14. Create a chart showing the mean (average) taxes paid, by accused status. Did one group tend to pay higher taxes than the other?

States

15. Use a scatterplot to explore the relationship between the number of fatal accidents in a state and the population of the state. Comment on the pattern, if any, in the scatterplot.
16. Use a scatterplot to explore the relationship between the number of fatal accidents in a state and the mileage driven within the state. Comment on the pattern, if any, in the scatterplot.

Nielsen

17. Chart the mean (average) rating by network. Comment on how well each network did that week. (Refer to your work in Session 2.)

Session 4

One-Variable Descriptive Statistics

Objectives

In this session, you will learn how to do the following:

- Compute measures of location and dispersion for a variable
- Create a box-and-whiskers plot for a single variable
- Compute z-scores for all values of a variable

Computing One Summary Measure for a Variable

There are several measures of location (mean, median, mode, and percentiles), and of dispersion (range, variance, standard deviation, etc.) for a single variable. You can use Minitab to compute or generate these measures. We'll start with the mode of an *ordinal* variable.

File ➤ Open Worksheet... Select the file called **Student**. The variables in this file are student responses to a first-day-of-class survey.

One of the variables in the file is called `Drive`. This variable represents students' responses to the question, "How would you rate yourself as a driver?" The answer codes are as follows:

AA = Above average
A = Average
BA = Below average

Stat ➤ Tables ➤ Tally Select the variable `Drive`, and click **OK**. Look at the results. ***What was the modal response? What, if anything, strikes you about this frequency distribution? How many students are in the "middle?" Is there anything peculiar about these students' view of "average?"***

Before continuing, **Edit your last dialog**, and check off **Percents, Cumulative Counts**, and **Cumulative Percents**. You should see this tally:

Summary Statistics for Discrete Variables

Drive	Count	CumCnt	Percent	CumPct
AA	104	104	47.71	47.71
A	106	210	48.62	96.33
BA	8	218	3.67	100.00
N=	218			
*=	1			

What does each of the resulting columns tell you?

`Drive` is a qualitative variable with three possible values. Some categorical variables have only two values, and are known as *binary* variables. Gender, for instance, is binary. In this dataset, there are two variables representing a student's sex. The first, which you have seen in earlier sessions, is called `Gender`, and is a Minitab text variable, assuming values of "F" and "M." The second is called `Female`, and is a numeric variable equal to 0 for men and 1 for women. If we wanted to know the *proportion* of women in the sample, we could tally either of the worksheet variables. Alternatively, we could compute the mean of `Female`. By summing all of the 0's and 1's, we would find the number of women; dividing by n would yield the sample proportion.

Calc ➤ Column Statistics... In this dialog box, choose the radio button for **Mean**, and select the variable `Female` as the input variable. Click **OK**.

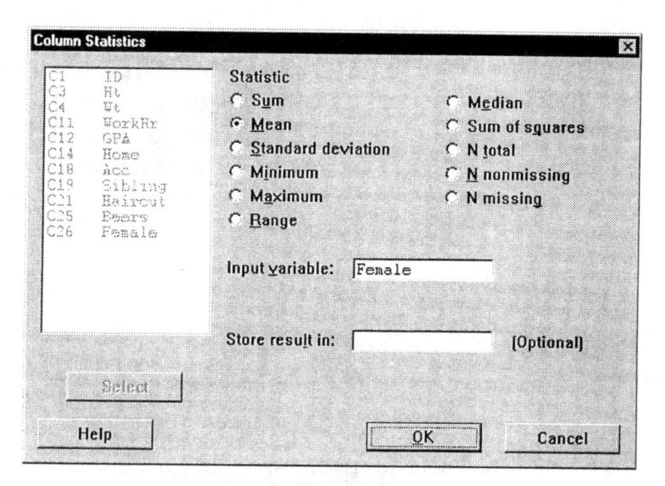

Now let's move on to a *quantitative* variable: the number of brothers and sisters the student has. The variable is called `Sibling`.

Use the **Tally** command to find the *mode* of the variable, `Sibling`. (You should de-select **Cumulative Counts** and **Cumulative Percents** in the **Tally** dialog.)

Now, find the other measures of central tendency. There are two ways to do this. If we simply want to know the mean or the median, we can compute it alone with the Column Stats command.

Calc ➤ Column Statistics... As you just did for `Female`, find the mean for `Sibling`.

Now look in the Session Window, and you will see the Mean number of siblings per student. ***Do you think that anyone in the class had this number of siblings? If not, then in what sense does the mean describe the variable? Would you prefer to Tally these results for a description of the class' answers?***

Computing Several Summary Measures

Suppose you wanted to know both the mean and standard deviation. The **Column Statistics** command will only compute one statistic at a time. To get several summary measures at one time, we use a different command.

Stat ➤ Basic Statistics ➤ Display Descriptive Statistics... Select the variables **HT** (height in inches) and **WT** (weight in pounds). You should see the output shown below. The output provides ten different descriptive statistics for each of the variables.

	N	MEAN	MEDIAN	TRMEAN	STDEV	SEMEAN
HT	84	68.143	68.000	68.184	3.973	0.434
WT	84	158.82	152.50	156.92	37.01	4.04

	MIN	MAX	Q1	Q3
HT	60.000	77.000	65.000	71.000
WT	102.00	265.00	130.00	179.75

Specifically, the command provides these summary measures:

N Number of observations in the sample for this variable

Mean The sample mean, or $\bar{x} = \frac{\sum x}{n}$

Median The sample median (50th percentile)

TrMean The 10% "trimmed" sample mean, computed by omitting the highest and lowest 5% of the sample data.[1]

StdDev The sample standard deviation, or $s = \sqrt{\frac{\sum (x - \bar{x})^2}{n-1}}$

SEMean The standard error of the mean. This measure becomes important later in your course, and its interpretation should be held off until then. The formula for the standard error is $s_{\bar{x}} = s/\sqrt{n}$

Min The minimum observed value for the variable

Max The maximum observed value for the variable

Q1 The first quartile (25th percentile) for the variable

Q3 The third quartile (75th percentile) for the variable

Compare the mean, median, and trimmed mean for the two variables. ***Does either of the two appear to have some outliers skewing the distribution?*** Look at a **dotplot** of the two variables to check your thinking.

---

[1] If a faculty member computes your grade after dropping your highest and lowest scores, she is computing a trimmed mean.

The **Descriptive Statistics** command offers several graphical options which relate the summary statistics to the graphs you worked with in earlier labs. For example, let's take a closer look at the heights.

Stat ➤ Basic Statistics ➤ Display Descriptive Statistics... Now select only `HT`, and then click the **Graphs** button in the dialog. Another dialog appears listing graphs you can display in addition to the descriptive summary. Choose **Graphical Summary**, and click OK in both dialogs boxes.

In the graphical summary (see below), you get all of the summary measures except Trimmed Mean, but you also see a histogram with a symmetrical bell-shaped curved superimposed, a box-and-whiskers plot, and two "confidence interval" graphs. The additional information refers to topics which we will discuss in later labs. At this point in the course, though, you know what most of this output means.

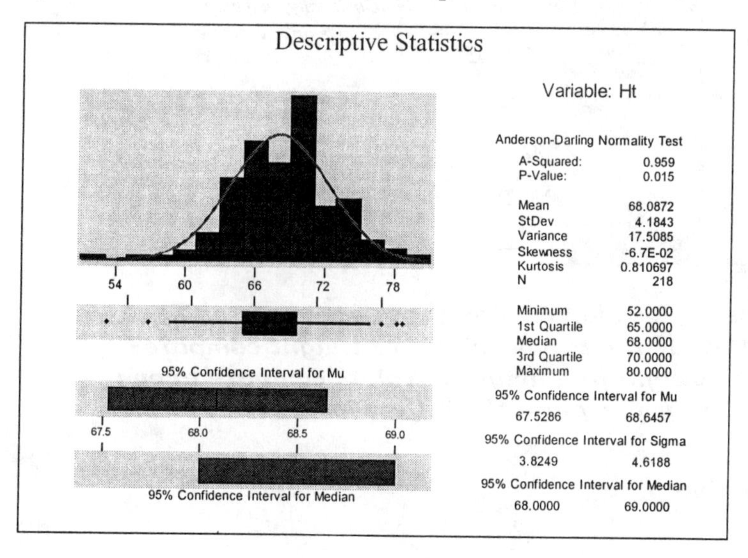

Generating a Box-and-Whiskers Plot

The descriptive statistics command generates the ***five-number summary*** for a variable (minimum, maximum, first and third quartiles, and median). A Box-and-Whiskers Plot visually displays the five number summary.[2] Additionally, it permits easy comparisons, as we will see.

Graph ➤ Boxplot Select `Ht` as the Y (measurement) variable, select `Gender` as the X variable, and click OK. In the resulting graph window, you'll see five-number summaries of the heights of men and women in the class. ***What does this graph suggest about the center and spread of the variable for the two sub-groups? Where are the median and quartiles in the graph?***

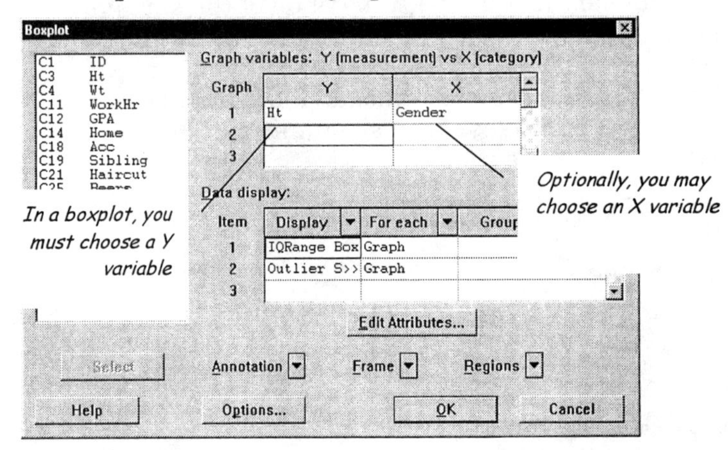

Try another boxplot for `Wt`, also using `Gender` as the X variable. ***How do the two boxplots for weight compare? How do the weight and height boxplots compare to one another? Can you explain the differences?***

Standardizing a Variable

Now open the file **Marathon.** This file contains the finish times for all wheelchair racers in the 1996 Boston Marathon.

---

[2] Actually, the whiskers in a Minitab boxplot may not extend to the minimum and maximum values. The lines project from the "box" at most a length of 1.5 times the interquartile range. Outliers are represented by asterisks.

Find the mean and median finish times. ***What do these two statistics suggest about the symmetry of the data?***

Since many of us don't know much about wheelchair racing or marathons, it may be difficult to know if a particular finish time is "good" or not. It is sometimes useful to *standardize* a variable, so as to express each value as a number of standard deviations above or below the mean. Such values are also known as *z-scores*.

Calc ➤ Standardize This command computes a z-score for each value of the Input variable and stores the computed values in a specified worksheet column. In this case, the input variable is `Minutes`. Move your cursor into the box marked **Store results in** and type `zscore`, and click OK.

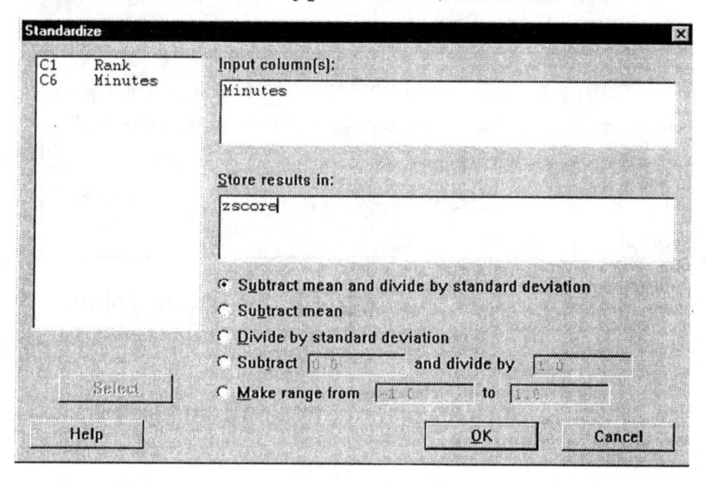

Now look at C7 in the Data Window. Since the racers are listed by finish rank, the first z-score value belongs to the winner of the race, whose finishing time was well below average. That's why his z-score is negative, indicating that his time was less than the mean.

Look at the z-scores of the top two racers. ***How does the difference between them compare to the difference between finishers #2 and #3? Between the last two finishers? What does this suggest about the difference between a ratio variable** (like* `Minutes`*) **and an ordinal variable** (like* `Rank`*)?*

Did anyone finish with a z-score of approximately 0? What does that indicate?

Moving On...

Now use the commands illustrated in this session to answer these questions. Where appropriate, indicate which statistics you computed, and why you chose to rely on them to draw a conclusion.

Student

1. What was the mean amount paid for a haircut?
2. What was the median amount paid for a haircut?

Colleges

This file contains tuition and other data from a 1994 survey of colleges and universities in the United States.

> In the Descriptive Statistics output you'll notice that the output includes both N (sample size) and N*. In a worksheet, * is the symbol for missing data. N* is the number of *missing observations* for a variable. In this instance, tuition data were not available for some of the schools in the sample.

3. In 1994, what was the average in-state tuition (**`Tuit_In`**) at US colleges? Out-of-state tuition? (**`Tuit_Out`**). Is it better to look at means or medians in this instance? Why?
4. Which varies more: in-state or out-of-state tuition? Why is that so?
5. Find your school in the Data Window (schools are listed alphabetically within state). In which quartile does your in-state tuition fall?

Output

This file contains data concerning industrial production in the U.S. from 1945–1996. Column 6 represents the degree to which the productive capacity of all US industries was being utilized. Column 7 has a comparable figure, just for manufacturers.

6. During the period in question, what was the mean utilization rate for all industrial production? What was the median? Describe the symmetry and shape of the distribution for this

variable. (Use the "graphical summary" in the descriptive statistics command.)

7. During the period in question, what was the mean utilization rate for manufacturing? What was the median? Describe the symmetry and shape of the distribution for this variable.

8. In terms of their standard deviations, which varied more: overall utilization or manufacturing utilization?

9. Comment on similarities and differences between the two variables.

Sleep

This file contains data about the sleeping patterns of different animal species.

10. Construct box-and-whiskers plots for `Lifespan` and `Sleep`. For each plot, explain what the "landmarks" on the plot tell you about each variable.

11. The mean and median for the Sleep variable are nearly the same (approximately 10.5 hours). How do the mean and median of Lifespan compare to each other? What accounts for the comparison?

12. According to the dataset, "Man" (row 34) has a life span of 100 years, and sleeps 8 hours per day. Determine, in terms of quartiles, where humans fall among the species for each of the two variables.

13. Sleep hours are divided into two types: "dreaming" and "non-dreaming" sleep. On average, do species spend more hours in dreaming sleep or non-dreaming sleep?

Water

These data concern water usage in 221 regional water districts in the United States for 1985 and 1990.

14. Column 17 (`to-cufr85`) is the total amount of fresh water used for consumption (drinking) in 1985. On average, how much drinking water did regions consume in 1985?

15. Column 34 (`percentcu`) is the percentage of all *fresh* water devoted to consumptive use (as opposed to irrigation, etc.) in

1985. What percentage of fresh water was consumed, on average, in water regions during 1985?

16. Which of the two distributions was more heavily skewed? Why was that variable less symmetric than the other?

Session 5

Two-Variable Descriptive Statistics

Objectives

In this session, you will learn how to do the following:

- Compute the coefficient of variation
- Compute measures of location and dispersion for subsamples of a variable
- Compute the covariance and correlation coefficient for two quantitative variables

Comparing Dispersion with the Coefficient of Variation

In the previous lab, you learned to compute descriptive measures for a variable, and to compare these measures for different variables. Often, the more interesting statistical questions require us to compare two sets of data, or to explore possible relationships between two variables. This session deals with techniques for making such comparisons and describing such relationships.

Comparing the means or medians of two variables is straightforward. On the other hand, when we compare the dispersion of two variables, it is sometimes helpful to take into account the magnitude of the individual data values. For instance, suppose we sampled the heights of mature maple trees and corn stalks. We could anticipate the standard deviation for the trees to be larger than that of the stalks, simply because the heights themselves are so much larger. What we need is a *relative measure* of dispersion. That is what the Coefficient of Variation (CV) is.

The CV is the standard deviation expressed as a percentage of the mean. Algebraically, it is:

$$CV = 100 \cdot \left(\frac{s}{\bar{x}} \right)$$

In Minitab, we can perform calculations like the CV and store the results either in columns or in ***constants***. Calculations which result in new *variables*, (i.e. which generate a value for each observation of an existing variable) are stored in columns; calculations which yield a single result are stored in constants. The CV formula will generate a **constant**.

Open the file called **Colleges**.

Stat ➤ Basic Statistics ➤ Display Descriptive Statistics... Select the variables `Tuit_In` and `Tuit_Out`. These values are different for state colleges and universities, but for private schools they are both the same. Not surprisingly, the mean for out-of-state tuition exceeds that for in-state.

But notice the standard deviations. ***Which variable varies more? Why is that so?*** The comparison is all the more interesting when we look at the Coefficient of Variation.

Calc ➤ Calculator Minitab represents constants as K1, K2, etc. In the box marked **Store result in variable**: type **K1**. Move your cursor into the **Expression** block.

In the Expression block, type the following (be careful to copy it precisely; see completed dialog below:)

```
100*stdev('Tuit_In')/mean('Tuit_In')
```

NOTE: When typing the apostrophes (single quotations marks), use the right-hand key with single and double quotation marks. *Do not use* the left-most key in the top row. You can also insert the expression 'Tuit_In' by double-clicking on it in the variable list.

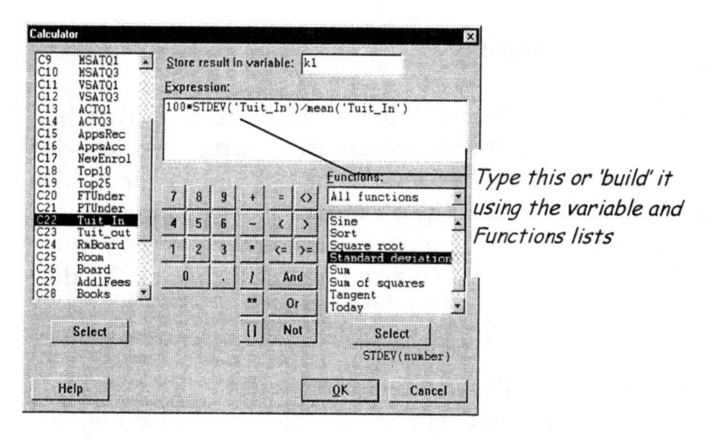

Click **OK**. You will not see any obvious result of your computation yet, but if you see an error message, try again.

Now repeat the computation for out-of-state tuition, storing the result in constant K2. K1 and K2 now contain the CV's for in- and out-of-state tuition, respectively.

To see the two constants, choose **Manip ➤ Display Data...** A dialog box will appear. Select K1 and K2, and click **OK**. Look in the Session Window for the results.[1] ***What do you notice about the degree of variation for these two variables? Does in-state tuition vary a little more or a lot more than out-of-state? Why does that happen?***

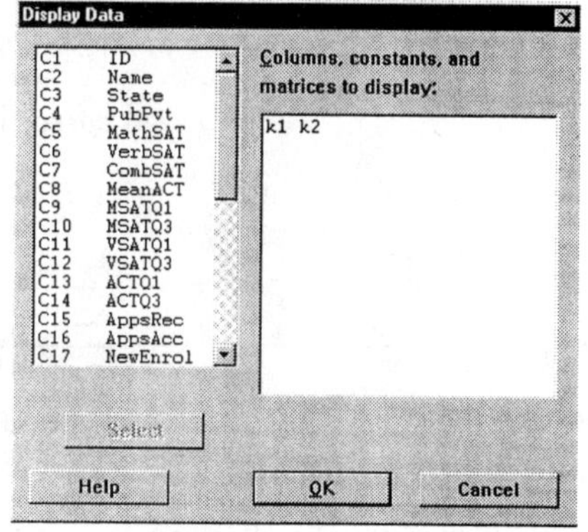

The standard deviations for the two variables were dollar amounts. ***What are the units for the Coefficient of Variation?***

---

[1] You can also find the current value of any stored constants by opening the **Info Window**. Open the Info Window, find the constants, and then *minimize* the window.

Descriptive Measures for Subsamples

Residency status is just one factor in determining tuition. Another important consideration is the difference between public and private institutions. We have a variable called "PubPvt" which equals 1 for public (state) schools, and 2 for private schools. In other words, the PubPvt column represents a qualitative attribute of the schools. We can compute separate descriptive measures for these two groups of institutions. To do so, we invoke the "By variable" option in the descriptive statistics command:

Stat ➤ Basic Statistics ➤ Display Descriptive Statistics Select both Tuit_in and Tuit_Out as the variables, and also check the small box next to **By Variable**. Then move the cursor into the box to the right of **By Variable**, and select PubPvt.

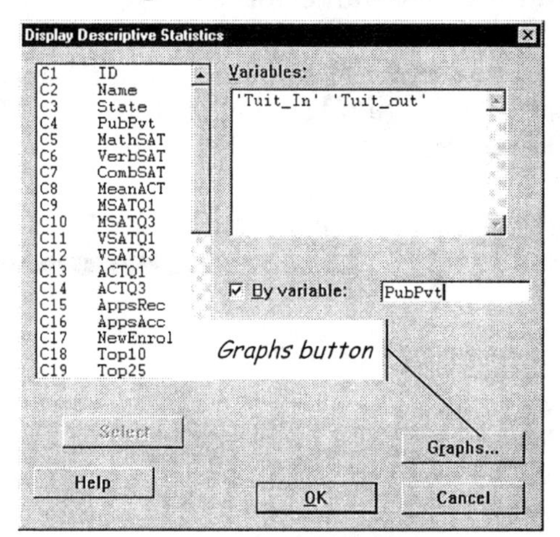

Before clicking **OK**, click on the **Graphs** button. In the **Graphs** dialog, select **Boxplot of Data.**

Two graph windows will open, one of which is shown below. ***What do the boxplots indicate about the comparative center and dispersion of tuitions for public and private colleges?***

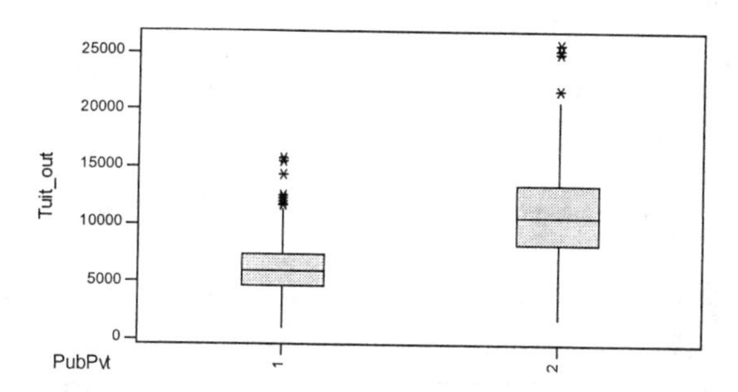

Look in the Session Window for the numerical results. These should look somewhat familiar, with one new twist. For `Tuit_Out`, two lines of output appear. The first refers to those sample observations with a `PubPvt` value of 1 (i.e., the State schools); the second refers to the Private school subsample. Take a moment to reconcile these statistics with the two boxplots. In the prior session, you determined the quartile "location" of your school. Do so again, using the quartiles for the public or private schools, as appropriate. ***Now where does your school rank?***

Measures of Association: Covariance and Correlation

We have just described a relationship between a quantitative variable (Tuition) and a qualitative variable (Public vs. Private). Sometimes, we may be interested in a possible relationship or association between two *quantitative* variables. For instance, in this dataset, we might expect that there is a relationship between the number of admissions applications a school receives (`AppsRec`) and the number of new students it accepts for admission (`AppsAcc`).

Stat ➤ Basic Statistics ➤ Display Descriptive Statistics Uncheck the **By variable** option. Select the variables `AppsRec` and `AppsAcc`. ***Do the boxplots and statistics make sense? How should the minimum and maximum values of these variables compare?***

Graph ➤ Plot Create a scatter plot for these variables, with `AppsAcc` on the Y axis, and `AppsRec` on the X axis. ***Do you see evidence of a relationship?***

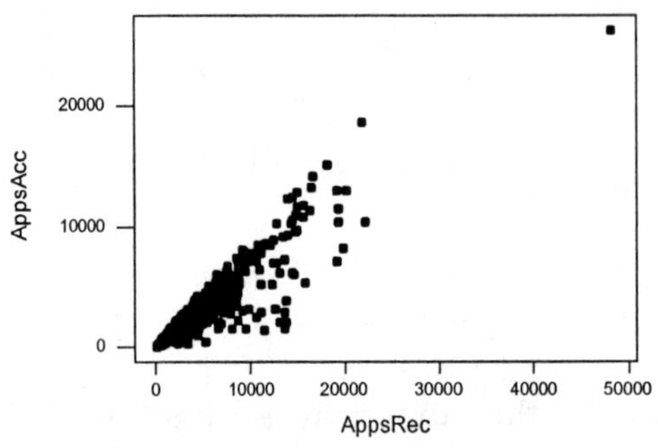

A graph like this one shows a fairly strong tendency for X and Y to "co-vary." In this instance, schools with higher X values also tend to have higher Y values. Given the meaning of the variables, this makes sense.

There are two common statistical measures of co-variation. They are the *covariance* and the *coefficient of correlation.* In both cases, they are computed using all available observations for a **pair of variables**. The formula for the sample covariance of two variables, x and y, is this:

$$\text{cov}_{xy} = \frac{\sum (x_i - \bar{x})(y_i - \bar{y})}{n-1}$$

The sample correlation coefficient[2] is:

$$r = \frac{\text{cov}_{xy}}{s_x s_y}$$

where:

s_x, s_y are the sample standard deviations of x and y, respectively.

Though these formulae may look daunting, it is easy to compute each of them.

---

[2] Formally, this is the Pearson Product Moment Correlation Coefficient, known by the symbol, *r*.

Stat ➤ Basic Statistics ➤ Covariance Select the variables **AppsRec** and **AppsAcc**, and click **OK**. You will see the following in the Session Window.

```
            AppsRec    AppsAcc
AppsRec    12545585
AppsAcc     7433971    5066399
```

In this triangular arrangement of values (known as a Variance-Covariance Matrix), the Covariance of **AppsRec** and **AppsAcc** appears at the intersection of the respective column and row. The covariance of **AppsRec** and **AppsAcc** is 7,433,971. What, then, are the other two numbers? They are the *sample variances* (s^2) of the two variables.

By definition, a correlation coefficient (symbolized r) always takes a value between -1 and +1. Absolute values near 1 are considered "strong" correlations; that is, the two variables have a strong tendency to vary together. Absolute values near 0 are weak correlations, indicating very little relationship or association between the two variables.

Stat ➤ Basic Statistics ➤ Correlation Again, select the variables **AppsRec** and **AppsAcc**, and click **OK**. You will find the correlation coefficient in the Session Window.

Variables can have strong sample correlations for a number of possible reasons. It may be that one is the cause of the other (or vice versa), that a third variable causes both of them, or that their observed association in this particular sample is merely a coincidence. As you will learn later in your course, correlation is an important tool in statistical reasoning, but we must never assume that correlation implies causation.

Moving On...

Use the commands and techniques presented in this session to answer the following questions. Explain your choice of statistics in responding to each question.

World90

This file contains economic and population data from 42 countries around the world. These questions focus on the distribution of Gross Domestic Product (GDP) in the countries.

1. Compare the means of C, I, and G (the proportion of GDP committed to consumption, investment, and government, respectively). Which is highest, on average?
2. Compare the mean and median for G. Why do they differ so?
3. Compare the coefficients of variation for C and for I. Which varies more: C or I? Why?
4. Compute the correlation coefficient for C and I. What does it tell you?

F500

This worksheet contains data about the 1996 Fortune 500 companies.

5. How strong an association exists between profit and revenue among these companies? (Hint: Find the correlation.)
6. One of the variables is "MktVal," representing the market value of the firm as of March 15, 1996. Presumably, several financial factors are related to market value. Which of these four seems to have the strongest relationship to market value: Revenue, Profits, Equity, or Growth (1985–95)? Explain your rationale, referring to statistical evidence.

Bev

This is the worksheet with data about the beverage industry.

7. If you have studied accounting, you may be familiar with the Current Ratio, and what it can indicate about the firm. What is the average current ratio in this sample of beverage industry firms?
8. In the entire sample, is there a relationship between the Current and Quick Ratios?
9. How do the descriptive measures for the Current and Quick ratios compare across the SIC subgroups? (Refer to Appendix A again for a full description of the SIC classification.) Suggest some possible reasons for the differences you observe.

Bodyfat

This dataset contains body measurements of 252 males.

10. What is the sample correlation coefficient between neck and chest circumference? Suggest some reasons underlying the strength of this correlation.
11. What is the sample correlation coefficient between biceps and forearm? Suggest some reasons underlying the strength of this correlation.
12. Which of the following variables is most closely related to bodyfat percentage (`FatPerc`): Age, Weight, Abdomen circumference, or Thigh circumference?

Salem

These are the data from Salem Village, Massachusetts in 1692. Refer to Session 4 for further description. Using appropriate descriptive and graphical techniques, compare the average taxes paid in the three groups listed below. In each case, decide whether you should compare means or medians, and explain why, as well as stating your conclusion.

13. Defenders vs. non-defenders
14. Accusers vs. non-accusers
15. Rev. Parris supporters vs. non-supporters

Sleep

This worksheet contains data about the sleep patterns of various mammal species. Refer back to Session 4 for more information.

16. Using appropriate descriptive and graphical techniques, how would you characterize the relationship (if any) between the amount of sleep a species requires and the mean weight of the species?
17. Using appropriate descriptive and graphical techniques, how would you characterize the relationship (if any) between the amount of sleep a species requires and the life span of the species?

Water

In Session 4, you described column 17, representing the total fresh water consumption in 1985. Column 33 contains comparable data for 1990.

18. Compare the means and medians for these columns, as well as boxplots. Did regions consume more or less water, on average, in 1990 than they did in 1985? What might explain the differences five years later?

19. Compare the coefficient of variation for each of the two variables. In which year were the regions more varied in their consumption patterns?

20. Construct a scatter plot of Fresh Water consumption in 1990 (c 33) versus the regional populations in that year (c18). Also, compute the correlation coefficient for the two variables. Is there evidence of a relationship between the two? Explain.

Session 6

Elementary Probability

Objectives

In this session, you will learn how to do the following:

- Simulate random sampling from a population
- Draw a random sample from a set of observations
- Manipulate worksheet data for analysis

Simulation

Thus far, all of our work with Minitab has relied on observed sets of data. Sometimes we will want to exploit the program's ability to *simulate* data which conforms to our own specifications.[1] In the case of experiments in classical probability, for instance, we can have Minitab simulate flipping a coin 10,000 times, or rolling a die 500 times.

A Classical Example

Imagine a game spinner with four equal quadrants, such as the one illustrated here. Suppose you were to record the results of 1000 spins. ***What do you expect the results to be?***

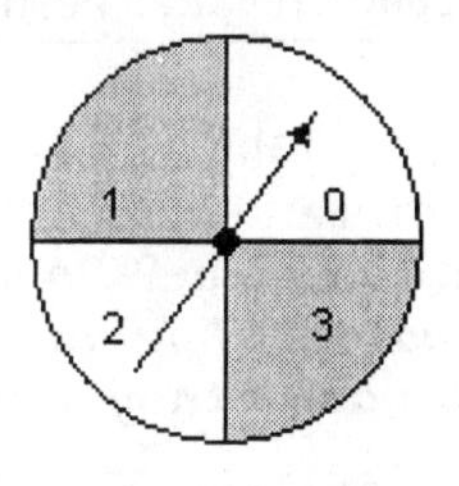

---

[1] More properly, these should be called *pseudo-random* samples, since the computer program follows an algorithm to generate them. In that sense, they are not literally random.

We can simulate 1000 spins of the spinner by having Minitab calculate some random data:

Calc ➤ Random Data ➤ Integer ... This command will calculate random integers to our specification. In particular, we must specify a range of allowable values, a number of observations, and a column location for the new data. Complete the dialog as shown here:

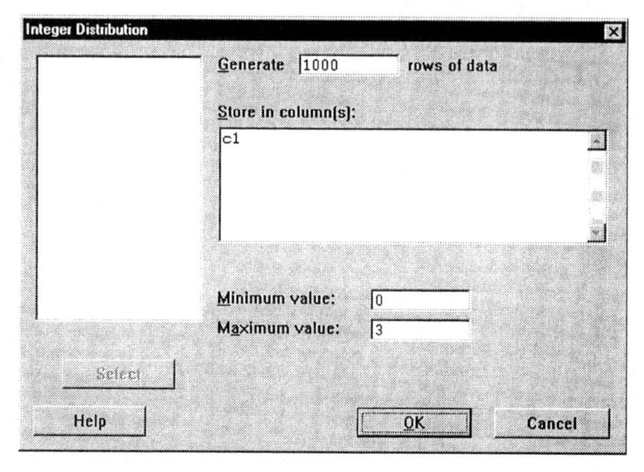

This will create a new variable, representing a random sample of 1000 spins. In Column 1 of the worksheet, you will see a list of values ranging from 0 through 3, simulating 1000 spins of the wheel.

> NOTE: Because these are random data, your data window will be unique. In other words, if you are working alongside a partner on another computer, her results will differ from yours.

Stat ➤ Tables ➤ Tally Tally the data in column 1, requesting counts and percents. What should the relative frequency be for each value? ***Do all of your results match the theoretical value exactly? To the extent that they differ, why don't they match?***

Recall that classical probabilities give us the long-run relative frequency of a value. Clearly, 1,000 spins does not adequately represent the "long-run," but this simulation may help you to understand what it means to say that the probability of spinning a "3" (or any other value) equals 0.25.

Observed Relative Frequency as Probability

As you know, many random events are not classical probability experiments, and we must rely on observed relative frequency. In this part of the session, we will direct our attention to the Census data, and focus on the chance that a randomly selected individual speaks a language other than English at home. The Census asked "Do you speak a language other than English at home?" These respondents gave three different answers: 0 indicates the individual did not answer or was under 5 years old; 1 indicates that the respondent spoke another language; and 2 indicates that the respondent spoke only English at home. ***Do you think these three answers occur with equal likelihood?***

File ➤ Open Worksheet Open the **Census90** worksheet.

Stat ➤ Tables ➤ Tally Tally the variable `Lang1`, selecting Counts and Percents. ***What do these relative frequencies indicate?***

Summary Statistics for Discrete Variables

Lang1	Count	Percent
0	66	6.72
1	64	6.52
2	852	86.76
N=	982	

If you were to choose one person from the 982 who answered this question, what is the probability that the person does speak a language other than English at home? Which answer are you most likely to get? Let's sample one person and see what happens.

Calc ➤ Random Data ➤ Sample from Columns... We want to sample one row from `Lang1`, and store the result in an empty column, which we will name `Sample`. Complete the dialog as shown on the next page:

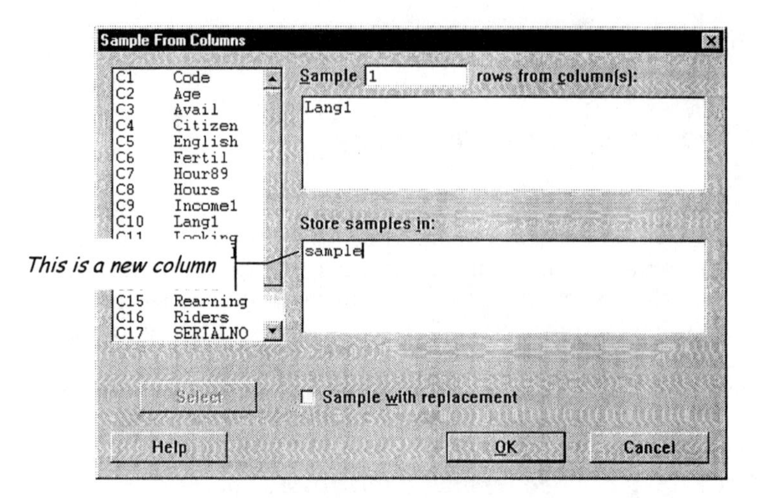

Look in the last column of the data window. There should be one cell in the column, containing a 0, 1, or 2. Which is it? ***What does it represent?*** From the tally we know that almost 87% of those responding speak only English. That notwithstanding, knowledge of the relative frequency is of limited value in predicting the outcome of asking one person. On the other hand, it does help us predict what pattern we'd see if we asked, say, 50 people. Now let's repeat this experiment 50 times:

- Edit your last dialog and request 50 rows of data.

- Tally these results (i.e., those of `Sample`). ***Do the relative frequencies match those of the entire dataset for Lang1? How do they compare?***

Handling Alphanumeric Data

In the prior example, the variable of interest was numeric. What if the variable is not represented numerically in the dataset?

- Open the file called **Colleges**. Imagine choosing one of these colleges at random. ***What's the chance of choosing a college from California?***

We could simply tally all of the data, and find out how many schools are in each state. That will give us a very long frequency table. Instead, let's see how to get a tally which just classifies all schools as being in California or elsewhere. To do so, we can first create a new binary variable, differentiating California from non-California schools. This requires several steps, as follows:

- In the data window, scroll right to a blank column. Label the new variable "**`CAL`**", as shown here. In Row 1 of that column, type `CA`, and then type **`Yes`** in the next cell to the right.

Colleges.MTW ***

	C29	C30	C31-T	C32-T	C33
↓	InstperS	GradRate	CAL		
1	10922	15	CA	Yes	
2	11935	*			
3	9584	39			

You type in these cells

This creates a "conversion table," establishing an equivalence between the categorical value "CA" and "Yes". Next we have Minitab create a new variable which equals "Yes" wherever **`State`** equals CA, and all other states are represented as missing.

- **Manip ➤ Code ➤ Use Conversion Table...** Complete the dialog box as shown below. The result of the dialog is to convert input values from **`State`** into a new variable called **`Calif`** using the conversion table we've just made.

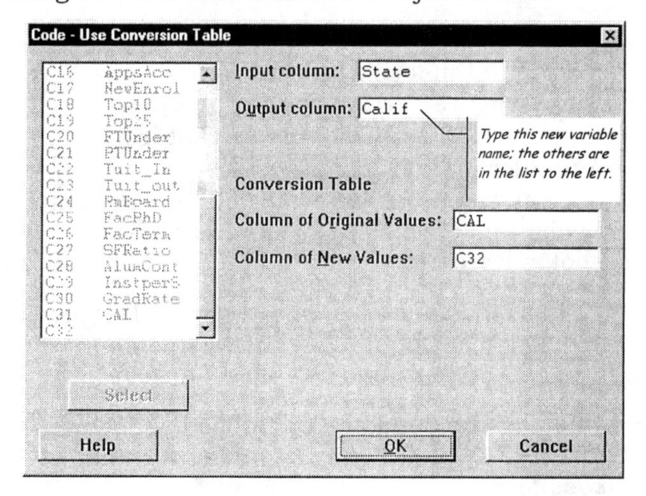

If you look in the Data Window, you'll find the new variable **`Calif`**. The first several rows in the **`Calif`** column are blank; scroll down to the California schools to see the effect of this command. This is nearly what we want; for further analysis, it will be helpful for all *non-California* schools to be coded as "No", rather than as missing data. Therefore, we now need to change all of the blanks to "No's". We do this with the Code command:

Manip ➤ Code ➤ Text to Text... The input column is `Calif`, and we want to store the result in the same place. We want to recode any blank, changing it to `No`.

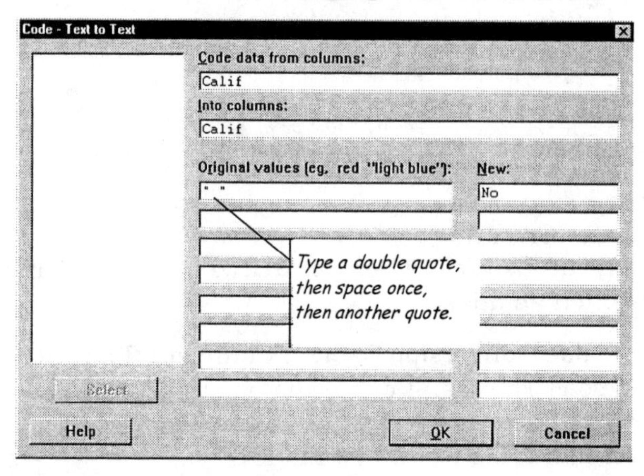

Now we have a column (`Calif`) which represents a *binary* variable: It equals `Yes` for schools in California, and `No` for all other schools. Now we can conveniently figure frequencies and relative frequencies. Similarly, we have a variable called `PubPvt` that equals 1 for public or state colleges and 2 for private schools.

In the language of elementary probability, let's define two events. Event "C" is that a randomly chosen school is in California. Event "Pv" is that a randomly chosen school is Private.

Stat ➤ Tables ➤ Cross tabulation The variables to select are `PubPvt` and `Calif`. Request **Counts** and **Total percents** to obtain frequencies and relative frequencies (see next page).

Look at the table in the Session Window, reproduced as follows. ***Can you interpret what you see? Does California have an unusual proportion of public colleges?***

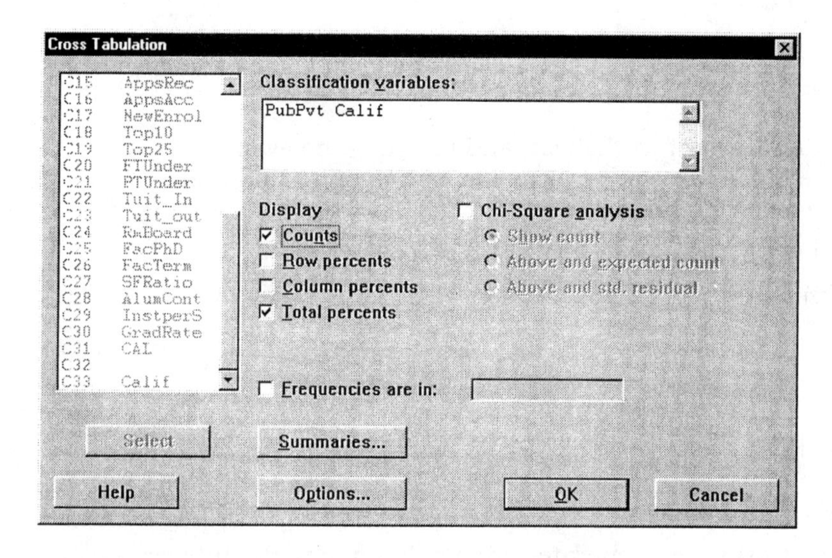

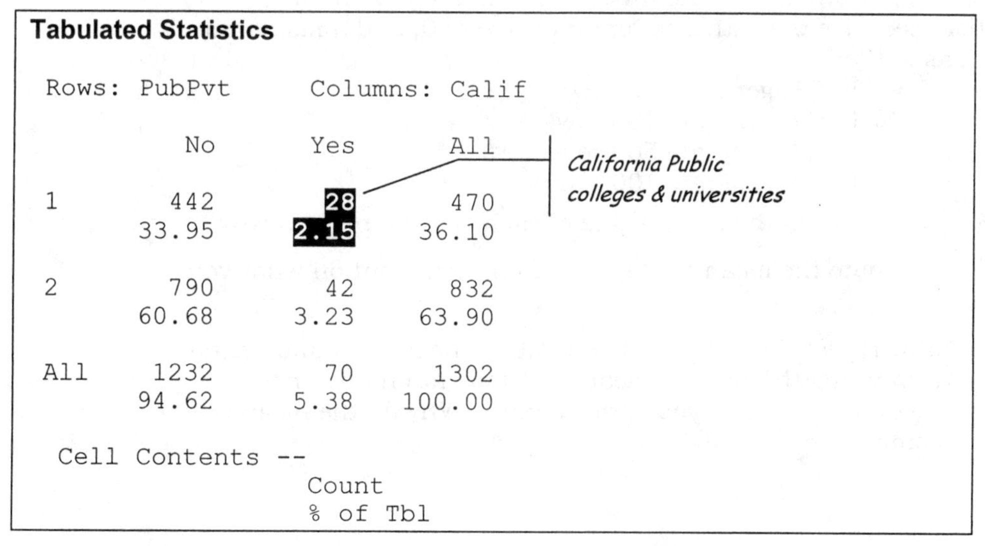

Tabulated Statistics

Rows: PubPvt   Columns: Calif

	No	Yes	All
1	442	**28**	470
	33.95	**2.15**	36.10
2	790	42	832
	60.68	3.23	63.90
All	1232	70	1302
	94.62	5.38	100.00

Cell Contents --
Count
% of Tbl

Moving On...

Within your current worksheet, and recalling the events previously defined, use the cross-tabulation command to find and comment on the following probabilities:

1. $P(C) = ?$
2. $P(Pv) = ?$
3. $P(C \cap Pv) = ?$
4. $P(C \cup Pv) = ?$
5. $P(Pv \mid C) = ?$

New Worksheet

In a fresh worksheet, generate random data using the Integer Distribution (see page 60) with a minimum value of 0, and a maximum value of 1, as follows:

- In C1, generate 10 rows
- In C2, generate 100 rows
- In C3, generate 500 rows
- In C4, generate 1000 rows

6. What should the mean value of each column be, and why?
7. Compute the mean for each column. Comment on what you find.
8. Now repeat the process of generating the four columns, and computing the column means. Comment on how these results compare to your prior results. Why do the means compare in this way?

Session 7

Discrete Probability Distributions

Objectives

In this session, you will learn how to do the following:

- Work with an observed discrete probability distribution
- Paste Session output into the Worksheet
- Compute the expected value of a distribution
- Compute binomial probabilities
- Compute Poisson probabilities

An Empirical Discrete Distribution

We already know how to summarize observed data; an *empirical distribution* is an observed relative frequency distribution which we intend to use to approximate the probabilities of a random variable. As an illustration, were we to randomly select a woman and ask how many children she has borne, we could regard that number to be a random variable. We will use the data in the Census90 file to illustrate.

> **File ➤ Open Worksheet** Open the worksheet file **Census90**. Recall that this contains 1990 Census questionnaire responses from Massachusetts residents.

In this file, we are interested primarily in the variable called `Fertil`, which is defined as the number children born to women 15 years of age and older. (See Appendix A for a detailed description.) Unfortunately, the dataset also includes men and preadolescent girls. Therefore, before analyzing the data, we need to extract a subsample.

To do this, we must take several steps. First, we'll extract the females, and then further extract only those over the age of 14.[1]

Manip ➤ Subset Worksheet... We'll open a new worksheet, containing all variables, but only for women. The name of the new worksheet doesn't matter, but we must specify a **Condition** as part of the dialog.

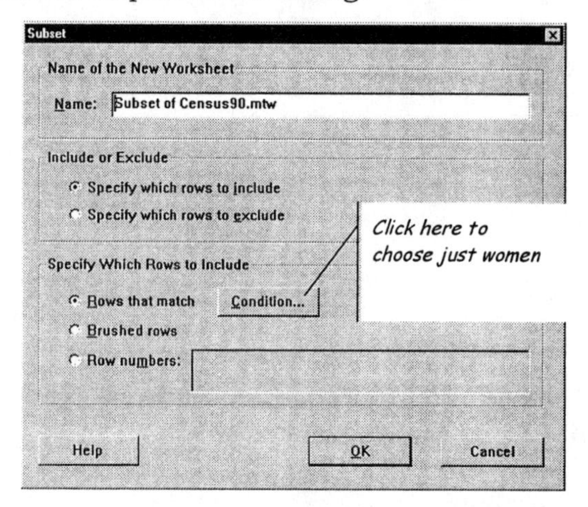

In the **Condition** dialog (below), type the expression **Sex=1**; This says that we want those observations for females. Click **OK**, and then **OK** in the main dialog.

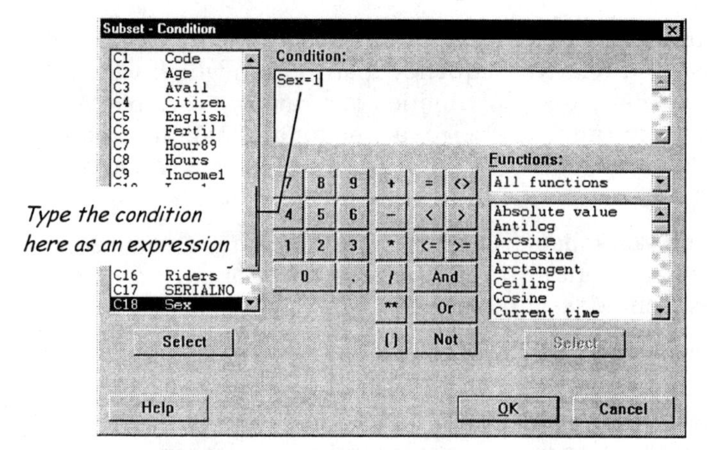

[1] Users of Minitab Release 11 should consult Appendix D.

This creates a new worksheet, containing only the responses of women. Now we need to cull out those women 15 years and older.

Manip ➤ Subset Worksheet... Using the same approach, we'll specify that we only want rows in which `Age > 14`.

In this worksheet, the column called `Fertil` is almost what we need. It contains responses from women over the age of 14, but its coding is potentially confusing. As noted in Appendix A, a value of 1 signifies 0 children, 2 signifies 1, and so on. Let's make the values easier to interpret:

Calc ➤ Calculator The variable is `Fertil`, and the expression is `Fertil - 1`. Click **OK**. This just subtracts one from each value, so that the variable does equal the number of children ever born to the woman responding.

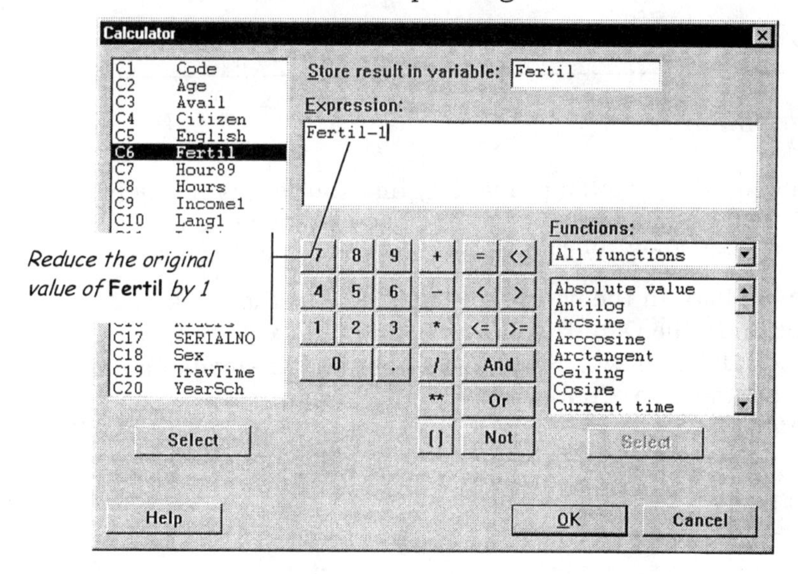

Stat ➤ Tables ➤ Tally Request **Counts** and **Percents** for the variable `Fertil`. ***In terms of probability, what do these percentages mean?***

Summary Statistics for Discrete Variables

Fertil	Count	Percent
0	115	28.97
1	71	17.88
2	88	22.17
3	61	15.37
4	38	9.57
5	10	2.52
6	5	1.26
7	4	1.01
8	1	0.25
9	2	0.50
10	1	0.25
12	1	0.25
N=	397	

Graphing a Distribution

It is often helpful to graph a probability distribution, typically by drawing a line at each possible value of X. The height of the line is proportional to the probability. We'll use the Chart command again.

Graph ➤ Chart. In the first graph, specify that the X variable is `Fertil`. In the Data Display area, change **Bar** to **Project**, and click OK. This will project a straight line, instead of a bar, at each data value. ***Comment on the shape of the distribution.***

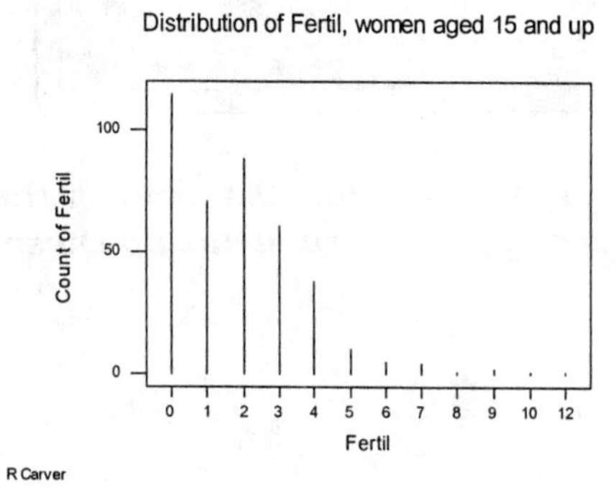

If we were to sample one woman at random, what's the most likely outcome? How many children, on average, did these women report? Which definition of "average" is most appropriate here?

Transferring Session Output to the Worksheet

Suppose we wanted to do further analysis of the Tally which we generated above. In particular, suppose you need to compute the expected value of the variable called `Fertil`. We can copy numbers from the Session Window, and paste them directly into the worksheet.

- Maximize[2] the Session Window, and scroll up so that you can see the entire tally. Position the cursor just to the left of the 0 in the `Fertil` column, and click and drag the mouse, highlighting the entire numerical portion of the tally. Release the mouse button so that only the tally itself is highlighted.

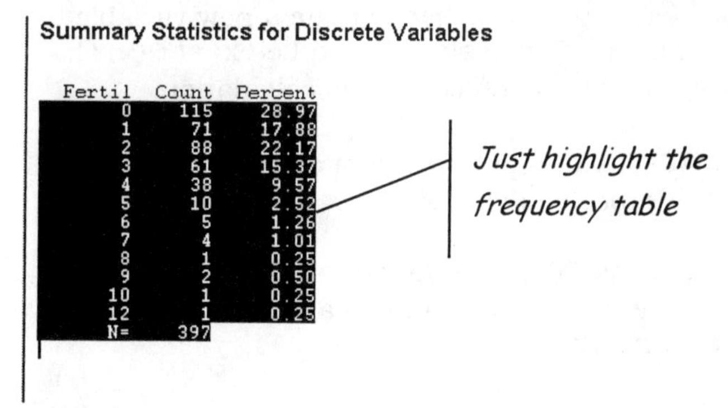

Summary Statistics for Discrete Variables

Fertil	Count	Percent
0	115	28.97
1	71	17.88
2	88	22.17
3	61	15.37
4	38	9.57
5	10	2.52
6	5	1.26
7	4	1.01
8	1	0.25
9	2	0.50
10	1	0.25
12	1	0.25
N=	397	

- After highlighting the relevant portion, select **Edit ➤ Copy**.

- Restore[3] the size of the Session Window to reveal the Data Window, and move the cursor to Row 1 of an empty column in the Data Window.[4]

---

[2] To *maximize* a window, click on the □ icon in the upper right corner of the screen (between – and ✕).

[3] To *restore* a maximized window to its former size, click on the 🗗 icon (between – and ✕).

[4] You may need to drag the open worksheet to find the scroll bars. Click on the bar at the top of the worksheet and drag the window slightly left and upward to reveal the scroll bars.

- **Edit ➤ Paste Cells** In the message dialog that appears, indicate that spaces should be treated as delimiters.

You should see the Frequency table (now a discrete distribution) appear in the worksheet. Label the three columns `X`, `f`, and `f(x)`.

Computing the Expected Value of X

The expected value of a random variable, X is the sum of the products of each value of X times its probability. The formula is:

$$E(x) = \mu = \sum xf(x)$$

The next two commands find that value.

- **Calc ➤ Calculator** Using the first and third columns in the newly created portion of the worksheet, create a new variable equal to X times f(x). (The **Expression** should be `'X'*'f(x)'`). Name the new column `Xf(x)`, reflecting the calculation.

- **Calc ➤ Column Statistics** Compute the **sum** of the new column. This is the *expected value* of X.

- **Calc ➤ Column Statistics** Compute the **mean** of `Fertil`, and compare it to the expected value you just calculated. ***Why do they relate in this way?***

A Theoretical Distribution: The Binomial

Some random variables arise out of processes which allow us to specify their distributions without empirical observation. Minitab can help us by simulating such variables or by computing their distributions. In this lab, we'll focus on the latter function.

- **File ➤ New** Create a new Minitab Worksheet.

- Into C1 of the blank worksheet, type the values 0 through 8. Label C1 `X` and C2 `f(x)`, as shown here:

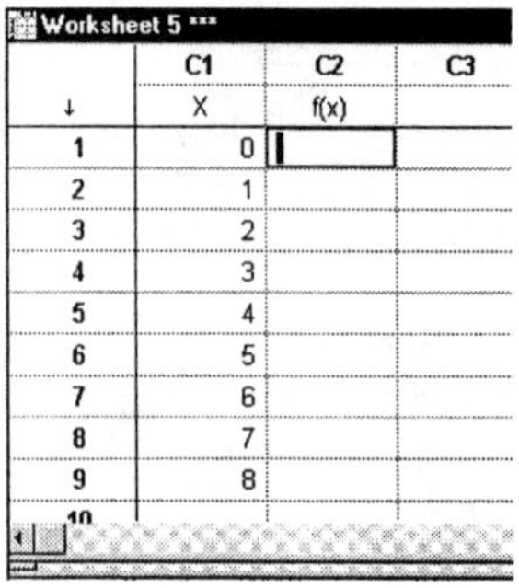

Worksheet 5 ***

↓	C1 X	C2 f(x)	C3
1	0		
2	1		
3	2		
4	3		
5	4		
6	5		
7	6		
8	7		
9	8		
10			

- **Calc ➢ Probability Distributions ➢ Binomial** In this dialog, specify that the **Number of trials** is 8, the **Probability of success** = .25, the **Input column** is `x`, and the **Optional storage** column is `f(x)`. Click **OK**. What does this command do in the Data Window?

- **Graph ➢ Plot** Your Y variable is `f(x)` and the X variable is `x`. In the **Data Display** portion of the dialog, change **Symbol** to **Project**, and click OK. ***Comment on shape of distribution; where is its peak? Does the peak have any special significance?***

- Title, label, and print this graph.

- **Calc ➢ Probability Distributions ➢ Binomial** Change the **Probability of success** to .4 and **Optional storage** to C3.

Compare C2 and C3. ***Comment on differences. How will the graph of C3 compare to that of C2?***

Another Theoretical Distribution: The Poisson

We can compute several common discrete distributions besides the Binomial. Let's look at one more. The ***Poisson*** distribution is often a useful model of events which occur over a fixed period of time. The distribution has but one parameter, and that is its mean. Using the same worksheet as for the binomial example, do the following:

- **Calc ➢ Probability Distributions ➢ Poisson** In this dialog, specify a **Mean** value of `2`, and **Input column** of `x`, and an **Optional storage** column of `f(x)`.

- Plot this variable as you did with the binomial, and print the graph. ***How do these graphs compare?***

Moving On...

Let's use what we have learned to try and (a) analyze an observed distribution and (b) see how well the Binomial or Poisson distribution serves as a model for the observed relative frequencies.

Student

Students were asked how many automobile accidents they had been involved in during the past two years. The variable called "Acc" records their answers. Perform these steps to answer the question below:

a) Tally the number of accidents, computing the relative frequencies.
b) Find the mean of this variable (equals the expected value)
c) In an empty column of the worksheet, type the values 0 through 9 (i.e., 0 in Row 1, 1 in Row 2, etc.)
d) Generate a Poisson Distribution with a mean equal to the mean number of accidents. The input column is the one you just typed, and the output column is the next one.
e) Type the computed relative frequencies (as proportions, not percentages) into the worksheet in the column adjacent to the Poisson probabilities.

1. Compare your tally of actual accidents to the Poisson distribution (either visually or graphically). Does the Poisson distribution appear to be a good model or approximation of the actual data?

Pennies

A professor has his students each flip 10 pennies, and record the number of heads. Each student repeats the experiment 30 times and then records the results in a worksheet.

2. Compare the actual observed results (in a graph or table) with the theoretical Binomial distribution with $n = 10$ trials and $p = 0.5$. Is the Binomial distribution a good model of what actually occurred when the students flipped the pennies? (Hint: Start by finding the mean of each column; how does the mean of a column relate to the probability of flipping x heads?)

Session 8

Probability Density Functions

Objectives

In this session, you will learn how to do the following:

- Create a column of "patterned data"
- Compute probabilities for any normal random variable
- Use normal curves to approximate other distributions

Continuous Random Variables

The prior session dealt exclusively with discrete random variables, that is, variables whose possible values can be listed (like 0, 1, 2, etc.). In contrast, some random variables are *continuous.* The defining characteristic of a continuous random variable is that, for any two arbitrarily selected values of the variable, there are an infinite number of other possible values between them. Such variables cannot be tabulated as can discrete variables, nor can a unique probability be assigned to each possible value. As such, we must think about probability in a different way in the context of continuous random variables.

Rather than constructing a probability distribution, as we did for discrete variables, we think in terms of a *probability density function* when dealing with a continuous random variable, X. We'll envision probability as being diffuse over the permissible range of X; sometimes the probability is "dense" near particular values, meaning that neighborhood of X values is relatively likely. The density function itself is difficult to interpret, but its *integral* (the area beneath the density function) represents probability.

Perhaps the most closely studied family of random variables is the *Normal Distribution.* We begin this session by considering several specific normal random variables.

Generating Normal Distributions

There are an infinite number of normally distributed random variables, each with its own mean and standard deviation. The mean and standard deviation, then, are the **parameters** of the distribution. If we know that X is normal, with mean μ and standard deviation, σ, we know all there is to know about X. Throughout this session, we'll denote a normal random variable as $X\sim N(\mu, \sigma)$. For example, $X\sim N(10,2)$ refers to a random variable X, which is normally distributed with a mean value of 10, and a standard deviation of 2.

The first task in this session will be to specify the density function for three different distributions, to see how the mean and standard deviation define a unique curve. Specifically, we'll generate values of the density function for a **standard normal variable,** $z\sim N(0,1)$, and two others: $X\sim N(1,1)$ and $X\sim N(0,3)$.

In the Data Window, label the first variable (column 1) **X**, label **C2** as **N01** (representing the standard normal variable), **C3** as **N11** and **C4** as **N03**.

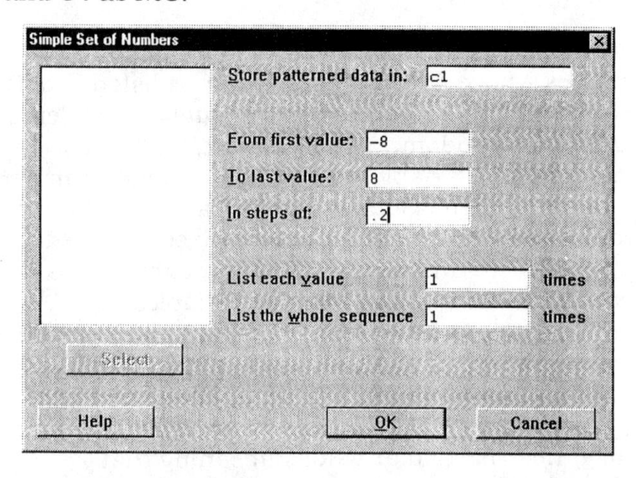

Calc ➤ Make Patterned Data ➤ Simple Set of Numbers In **C1**, we want to generate a column of values ranging from -8 to +8

with an increment of 0.2. We could type all of the values into the column, but this command does it for us.

Calc ➤ Probability Distributions ➤ Normal... This command will compute values of the normal density function. Specify a **Mean** = 0 and **Standard deviation** = 1. The input column is **X**, and the **Optional storage** column is **N01**. Now C2 contains density values for X~N(01).

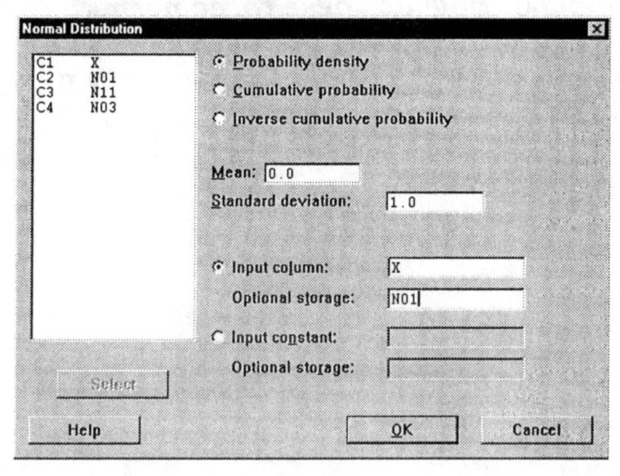

Edit ➤ Edit Last Dialog Change the **Mean** to 1, and the **Optional storage** column to N11.

Edit the last dialog once more, this time changing the mean to 0, the standard deviation to 3, and the optional storage column to N03.

If you were to graph these three normal variables, how would the graphs compare? Where would the curves be located on the number line? Which would be steepest and which would be flattest? Let's see:

Graph ➤ Plot In the main dialog box, specify the variables for three graphs. The Y variables are N01, N11, and N03. In each graph, type X in the X column. Before clicking **OK**, do this:

In the **Data display** portion of the dialog, change **Symbol** to **Connect.** This will draw short line segments between points, approximating a smooth curve.

Click **Edit Attributes** to assign different colors to the three graphs. Click **OK**.

Frame ➢ Multiple Graphs... Check **Overlay graphs on the same page**. Click **OK**, and then **OK** in the Plot dialog.

Title and place your name on the graph.

Look at the resulting graph. ***How do these three normal distributions compare to one another? How do the two distributions with a mean of 0 differ? How do the two with a standard deviation of 1 differ?***

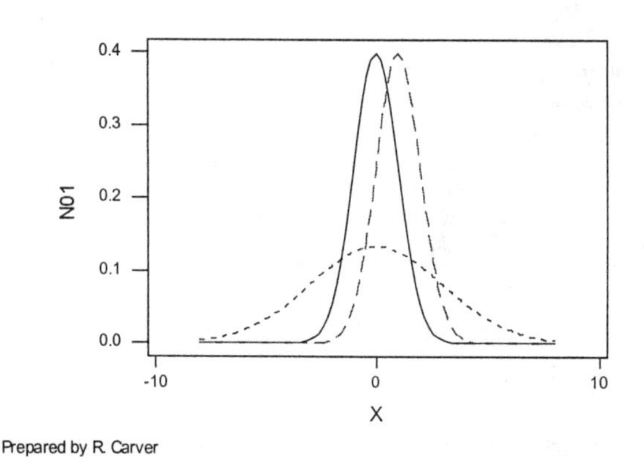

Print this graph.

Finding Areas Under a Normal Curve

We often need to compute the probability that a normal variate lies within a given range. We generally convert a variable to the standard normal variable, z, consult a table of values, and then manipulate the areas to find the probability. We can find these probabilities easily.

Calc ➢ Probability Distributions ➢ Normal... At the top of the dialog box, click **Cumulative Probability**. The values generated represent the area under the curve from -∞ to the specified value in the **Input Column**. Set the **Mean** = 0 and **Standard**

Deviation = `1`. The input column is still `X`. Type `CumN01` in the **Optional storage** block to generate the cumulative probabilities for the standard normal variable, z.

Plot this new column (`CUMN01`) versus `X`.

> You will see a warning message that you have changed the number of graphs. In your prior Plot dialog, you specified 3 graphs on one set of axes. Minitab is alerting you that it is about to discard your earlier instruction concerning multiple graphs. Just click **OK**.

This is a less-than ogive for the standard normal variable. ***What is notable about the shape of CUMN01? What is the special significance of the point (0,0.5)?***

Now suppose we want to find P (-2.5 < z < 1). We could scroll through Column 5 to locate the probabilities, or we could request them directly, as follows.

In empty column (C6), type the two values `-2.5` and `1`.

Calc ➤ Probability Distributions ➤ Normal You only need to change the **Input column** to `C6` and the **Optional storage** to `C7`. After clicking **OK**, you'll see this in the worksheet:

C4	C5	C6	C7	C
N03	CumN01			
.003799	0.00000	-2.5	0.006210	
.004528	0.00000	1.0	0.841345	
.005373	0.00000			
.006347	0.00000			
.007465	0.00000			

Subtract the two values in C7 for the result. This approach works for any normally distributed random variable. Suppose X is Normal with a mean of 500 and a standard deviation of 100. Let's find P(500 <x <600)

Type `500` and `600` into top two cells of C6

Edit the last command dialog again. Change the **Mean** to `500`, the **Standard Deviation** to `100`. We can retain columns 6 and 7 for input and output. Once again, subtract the two result values.

What is P(X > 600)? P(X < 300)?

Normal Curves as Models

One reason that the normal distribution is so important is that it can serve as a close approximation to a variety of other distributions. For example, binomial experiments with many trials are approximately normal. Let's try an example of a binomial variable with 100 trials, and P(success) = .20.

Calc ➤ Make Patterned Data ➤ Simple Set of Numbers In C8, generate rows from 0 to 100 with an increment of 1.

Calc ➤ Probability Distributions ➤ Binomial Set the number of trials = 100, p(success) = .2 The **Input column** is `C8`, and **Optional storage** is `C9`. C9 now contains the Binomial probability values for the number of successes in 100 trials.

Graph ➤ Plot C9 vs. C8

Do you see that this distribution could be approximated by a normal distribution? The question is, *which* normal distribution in particular? Since n = 100 and p = .20, the mean and standard deviation of the binomial variable are 20 and 4.[1] Let's generate a normal curve with those parameters.

Calc ➤ Probability Distributions ➤ Normal Click **Probability Density**. The **Mean** = `20`, **Standard Deviation** = `4`. The **Input column** is `C8`, **Optional storage** is `C10` (which is currently empty).

Plot C10 vs. C8 on the same axes as C9 vs. C8 [**Frame ➤ Multiple Graphs**]. Use different colors for the two lines [**Edit Attributes**] ***Would you say the two curves are approximately the same?***

---

[1] For a binomial X, $E(X) = \mu = np$. Here, that's (100)(.20) = 20. Similarly, the standard deviation is $\sigma = \sqrt{np(1-p)} = \sqrt{(100)(.20)(80)} = \sqrt{16} = 4$.

The normal curve is often a good approximation of real-world observed data. Let's consider two examples.

- **Open** the file **Paworld**, which contains annual economic and demographic data from 42 countries.

- **Graph ➤ Histogram** Ask for two graphs: one of C (% of GDP consumed) and one of Y (Real per capita GDP relative to the United States). ***Does a normal distribution approximate either of these histograms?***

- **Stat ➤ Basic Statistics ➤ Display Descriptive Statistics** Select the same two variables (C and Y). Click on **Graphs**, and select **Histogram of data, with normal curve**, and then execute the command. The parameters of superimposed normal curves are the sample mean and standard deviation for each variable.

- **Window ➤ Manage Graphs** In the dialog, there is a list of graphs. Highlight `PAWorld.MTW: Nhist of C.` Then, holding down the Shift key, move the mouse arrow down to the next graph and click the left mouse button, highlighting the two graph titles.

- Click **Tile**, and **Done**. You will see both graphs displayed side-by-side.

In your judgment, how closely does the normal curve approximate each histogram?

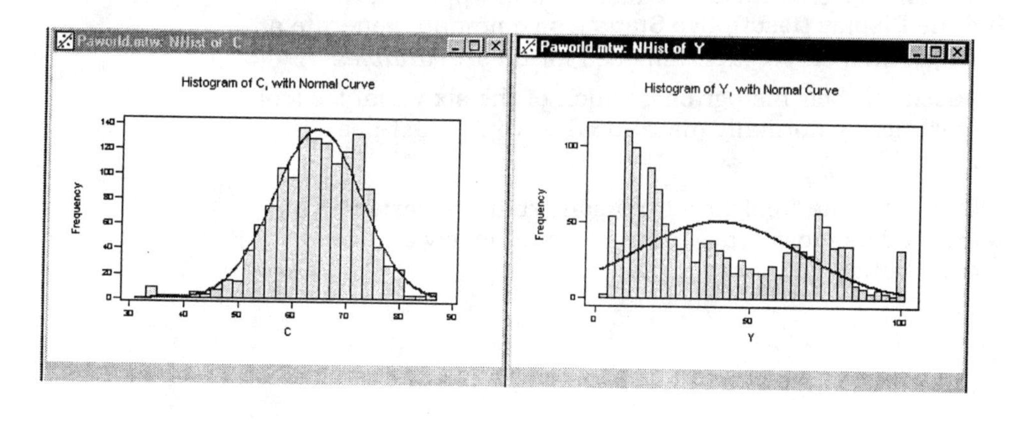

Moving On...

Worksheet 1 (containing the simulated data from the first part of the session)

1. Use what you have learned to compute the following probabilities for a normal random variable with a mean of 8 and a standard deviation of 2.5:
    - $P(7 < X < 8.5)$
    - $P(9 < X < 10)$
    - $P(X > 4)$
    - $P(X < 4)$
    - $P(X > 10)$

2. Into a blank worksheet column, generate a simple set of numbers from 0 to 160. In the adjacent column, generate Binomial probabilities for a binomial distribution with parameters $n = 160$ and $p = 0.4$. In the next column, compute the appropriate normal probability densities. Construct a graph to compare the binomial and normal probabilities; comment on the comparison.

Output

This file contains monthly data about the industrial output of the United States for many years. The first column contains the date, and the next six contain specific variables described in Appendix A.

Using the **Display Descriptive Statistics** command, generate a histogram with normal curve superimposed for *all six variables*.

3. Based on their histograms, which of the six variables looks most nearly normally distributed to you? Least nearly normal?

4. Suggest some "real world" reasons that the variable you selected as most nearly normal would follow a normal distribution.

Bodyfat

This file contains body measurements of 252 men. Using the same technique described for the **Output** dataset, investigate these variables:

- FatPerc
- Age
- Weight
- Neck
- Biceps

5. Based on their histograms, which of the variables looks most nearly normally distributed to you? Least nearly normal?
6. Suggest some "real world" reasons that the variable you selected as most nearly normal would follow a normal distribution.
7. For the neck measurement variable, what are the parameters (mean and standard deviation) of a normal curve, which closely approximates the observed data?
8. Use Minitab's cumulative normal command and the normal distribution just identified to estimate the percentage of men with neck measurements between 29 and 35 cm.

Water

These data concern water usage in 221 regional water districts in the United States for 1985 and 1990. Compare the normal distribution as a model for C17 and C34 (you investigated these variables earlier in Session 4).

9. Which one is more closely modeled as a normal variable?
10. What are the parameters of the normal distribution which closely fits the variable `Percentcu` (C34)?
11. What concerns might you have in modeling `Percentcu` with a normal curve?

Mft

This worksheet holds scores of 137 students on a Major Field Test, as well as their GPA's and SAT verbal and math scores.

12. Identify the parameters of a normal distribution, which closely approximates the Math scores of these students.

13. Use Minitab's cumulative normal command and the distribution you have identified to estimate how many of the 137 students scored more than 59 on the Math SAT.

14. We know from the Descriptive Statistics command that the Third Quartile (75 percentile) for Math was 59. How can we reconcile your previous answer and this information?

Session 9

Sampling Distributions

Objectives

In this session, you will learn how to do the following:

- Simulate random sampling from a known population
- Use simulation to illustrate the Central Limit Theorem

What is a Sampling Distribution?

Every random variable has a probability distribution or a probability density function. One special class of random variables is that of *statistics computed from random samples.*

How can a statistic be a random variable? Consider a statistic like the sample mean, $\bar{x}$. In a particular sample, $\bar{x}$ depends on the n values in the sample; a different sample would potentially have different values, probably resulting in a different mean. Thus, $\bar{x}$ is a quantity that varies from sample to sample, due to the luck of random sampling. In other words, it's a quantitative random variable.

. Every random variable has a distribution with shape, center and dispersion. The term *sampling distribution* refers to the distribution of a sample statistic. In this lab, we'll simulate drawing many random samples from populations whose distributions are known, and see how the sample statistics vary from sample to sample.

Sampling from a Normal Population

We start by considering a large sample from a population known to be normally distributed, with $\mu = 500$ and $\sigma = 100$.

Calc ➤ Random Data ➤ Normal Generate 200 rows in C1, **Mean** = 500, **Standard deviation** = 100, as shown here:

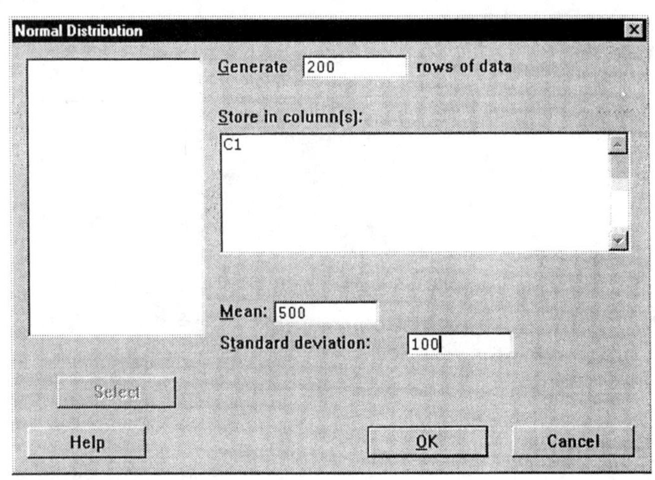

Look at C1 in your Data Window. Remember that it's a random sample, different from your neighbors' and different still from another sample you might have drawn. The question is, how much different?

Since the mean of the population is 500, it is reasonable to expect the mean of this first column to be near 500. It may or may not be "very" close, but the result of one simulation doesn't tell us much. To get a feel for the randomness of $\bar{x}$, we need to consider many samples.

Let's think of C1 as containing the first observation (X_1) of 200 simulated random samples. Let's see what happens if we simulate 200 samples, each with n = 50 observations. We'll proceed to generate 200 rows in the next 49 columns of the worksheet. Then we can compute the sample means for each of the 200 samples, create a column containing all 200 sample means, and then make some comparisons among them.[1]

---

[1] It may seem more "natural" to treat the columns as samples, since we usually think of observations as occupying rows. Because of the simplicity of computing *and* storing the means of rows (but not columns) in Minitab, we'll proceed as described in this session.

Edit ➤ Edit last dialog... In the **Store in column(s)** box, change C1 to C2-C50.

Move to the Data Window, and label C51 as **Mean**.

Calc ➤ Row Statistics Select **Mean** of C1 - C50, and store the result in C51, as shown here:

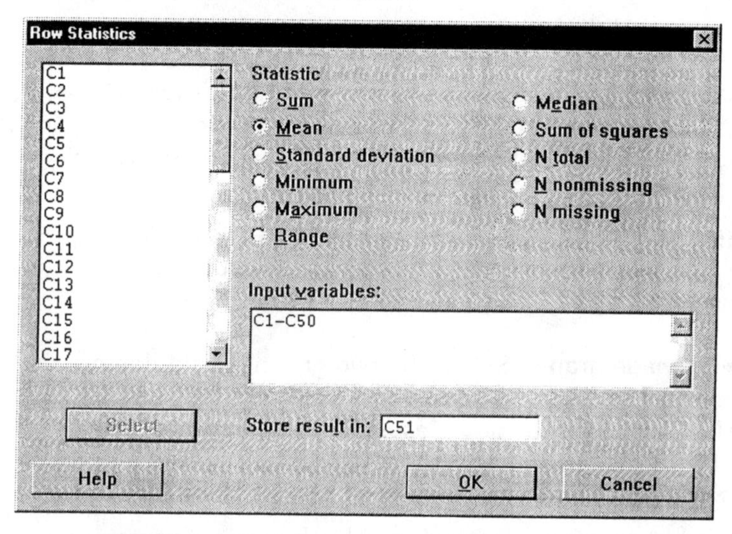

C51 now contains the means of all 200 samples; we can consider it a random variable, since each sample mean is different due to the chance involved in sampling. ***What should the mean of all of these sample means be?***

Stat ➤ Basic Statistics ➤ Display Descriptive Statistics Select C51.

Despite the fact that your random samples are unique and individually unpredictable, we can predict that the mean of C51 will be very nearly 500. This is a key reason that we study sampling distributions. We can make very specific predictions about the sample mean in repeated sampling, even though we cannot do so for one sample.

How much do the sample means vary around 500? Recall that in a random sample from an infinite population, *the standard error of the mean* is given by this formula:

$$\sigma_{\bar{x}} = \frac{\sigma}{\sqrt{n}}$$

In this case, σ = 100 and n = 50. So here,

$$\sigma_{\bar{x}} = \frac{100}{\sqrt{50}} = \frac{100}{7.071} = 14.14$$

Now find the standard deviation of C51 in the Session Window. Is it approximately 14.14? Remember that the standard error is the theoretical standard deviation of all possible values of $\bar{x}$, and the standard deviation of C51 represents only 200 of those samples.

So far we've looked at the center and spread of the distribution of $\bar{x}$; what about its shape?

Edit ➤ Edit last dialog... Choose C50 and C51. Before clicking **OK**, click on **Graphs**, and choose **Histogram of data, with normal curve.**

Window ➤ Manage graphs Select the two graphs (hold down Shift key), and **Tile** them, so they appear side-by-side as shown below. ***How would you describe the shapes of the graphs? What do you notice about their respective centers and spread? (Look closely at the horizontal axis).***

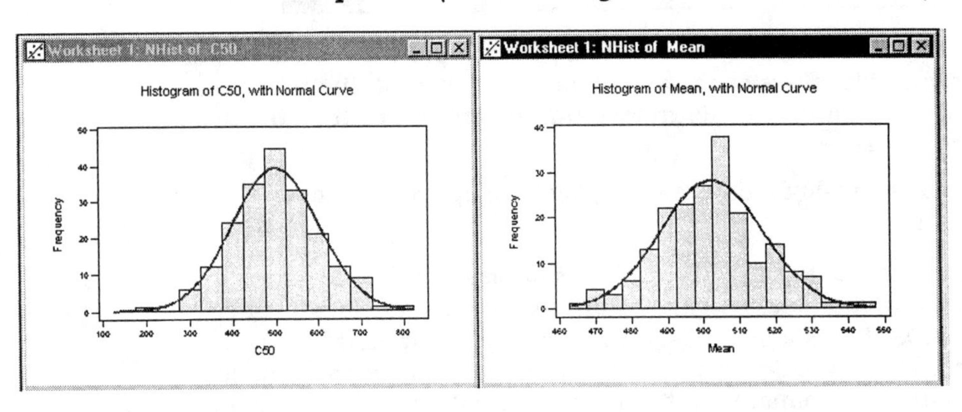

Print the two graphs.

Central Limit Theorem

The histogram of C51 was roughly normal, describing the means of many samples from a normal population. That may seem reasonable—the means of samples from a normal population are themselves normal. But what about samples from non-normal populations?

According to the Central Limit Theorem, the distributions of sample means approach a normal curve as n grows large, *regardless of the shape of the parent population.* To illustrate, let's take 200 samples from a uniform population ranging from 0 to 100. In a *Uniform* population with a minimum value of a and a maximum value of b, the mean is found by:

$$E(x) = \mu = \frac{(a+b)}{2}$$

In this population, that works out to a mean value of 50. Furthermore, we can compute the population variance as:

$$Var(x) = \sigma^2 = \frac{(b-a)^2}{12}$$

In this population, the variance is 833.33, and therefore the standard deviation is σ = 28.8675. Our samples will once again have n=50; according to the Central Limit Theorem, the standard error of the mean in such samples will be $28.8675/\sqrt{50} = 4.08$. Thus, the Central Limit Theorem predicts that the means all possible of 50 observation samples from this population will follow a normal distribution whose mean is 50 and standard error is 4.08. Let's see how well the theorem predicts the results of this simulated experiment.

- **Calc ➤ Random Data ➤ Uniform** Generate `200` rows and store them in `C1-C50`. Specify a **Lower endpoint** of `0` and an **Upper endpoint** of `100`.

- **Calc ➤ Row Statistics** Once again, select **Mean** of `C1-C50`, and **Store the results** in `C51`.

- **Stat ➤ Basic Statistics ➤ Display Descriptive Statistics** Select `C50` and `C51` once again, creating the histograms as in the earlier simulation. ***Describe the center, shape, and spread of each graph.*** Look in the Session Window to find the mean and standard deviation of the two columns. ***Do the mean and***

standard error of `C51 (Mean)` ***match the theoretical values predicted by the Central Limit Theorem?***

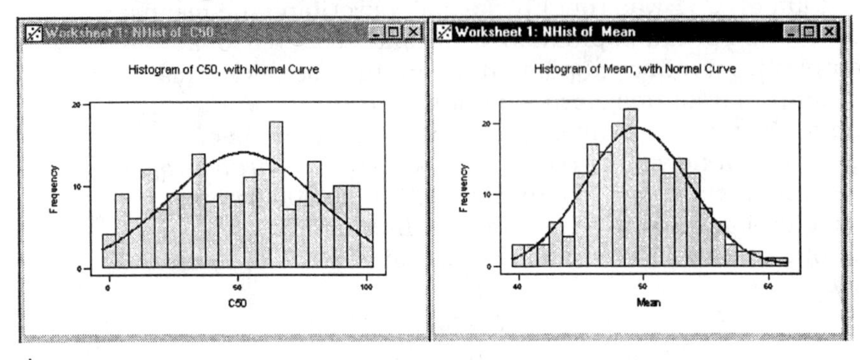

Look closely at your two graphs (mine are shown here). ***What similarities do you see between your graphs and these? What differences? How do you explain the similarities and differences?***

Sampling Distribution of the Proportion

The examples thus far have simulated samples of a quantitative random variable. Not all variables are quantitative. The Central Limit Theorem and the concept of a sampling distribution also apply to qualitative random variables, with three differences. First, we are not concerned with the mean of the random variable, but with the *proportion (p)* of times that a particular outcome is observed. Second, we need to change our working definition of a "large sample." The standard rule of thumb is that n is "large" if both $n \cdot p > 5$ and $n(1 - p) > 5$. Third, the formula for the standard error becomes:

$$\sigma_{\bar{p}} = \sqrt{\frac{p(1-p)}{n}}$$

To illustrate, we'll generate more random data. Recall what you learned about Binomial experiments as a series of n independent trials of a process generating success or failure with constant probability, p, of success. Such a process is known as a *Bernoulli trial.* We'll construct 200 more samples, each consisting of 50 Bernoulli trials:

> **Calc ➤ Random Data ➤ Bernoulli** Generate 200 rows in columns 1 through 50. Specify that **Probability of success** equals `0.3`.

This creates 50 columns of 0's and 1's, where "1" represents a success. By finding the mean of each row, we'll be calculating the relative frequency of successes in each of our simulated samples, also known as the *sample proportion,* $\bar{p}$.

Calc ➤ Row Statistics Once again, select **Mean** of `C1 - C50`, and **Store the results** in `C51`.

Now column 51 contains 200 sample proportions. According to the Central Limit Theorem, they should follow an approximate normal distribution with a mean of 0.3, and a standard error of

$$\sigma_{\bar{p}} = \sqrt{\frac{p(1-p)}{n}} = \sqrt{\frac{(.3)(.7)}{50}} = .0648$$

As we have in each of the simulations, compute and graph the descriptive statistics for columns 50 and 51. ***Comment on the descriptive statistics and the graphs you see.***

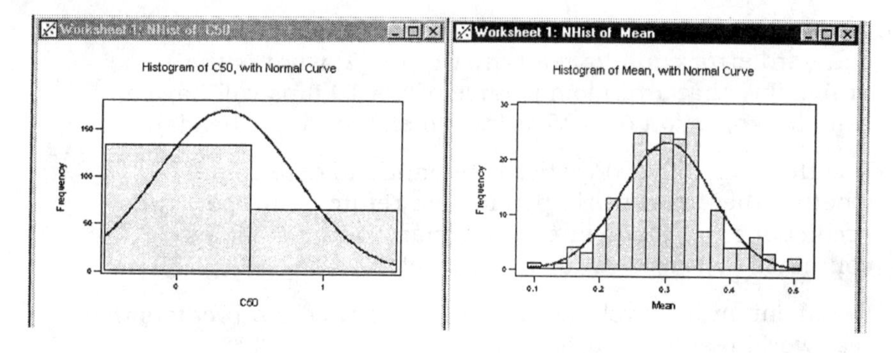

Moving On...

What happens when n is under 30? Does the Central Limit Theorem work for small samples too? Using the Uniform distribution, try generating only 20 columns of data as you did above, and examining the histogram of the sample means.

If your computer's memory is large enough, try larger samples as well. Generate 50 or 60 columns and see what happens.

1. This lab has used samples from two known distributions: one normal and one uniform. Minitab can generate random samples from a wide variety of other populations as well. Do a comparable demonstration using a uniform distribution

ranging from –10 to 10, and report on the distribution of sample means from 200 samples of $n = 50$.

Pennies

This file contains the results of 1,685 repeated binomial experiments, each one of which consisted of flipping a penny 10 times. We can think of each 10-flip repetition as a sample of $n = 10$ flips; this file summarizes 1,685 different samples.

In all, nearly 17,000 individual coin flips are represented in the file. Each column represents a different possible number of heads in the ten-flip experiment, and each row contains the results of one student's repetitions of the 10-flip experiment. Obviously, the average number of heads should be 5, since the theoretical proportion is $p = 0.5$.

2. According to the formula for the standard error of the sample proportion, what should the standard error be in this case (use $n = 10$, $p = .5$)?

3. Assuming a normal distribution, with a mean = 0.5 and a standard error equal to your answer to #2, what is the probability that a random sample of $n = 10$ flips will have a sample proportion of *0.25 or less*? (i.e., 2 or fewer heads)

4. Use the descriptive statistics commands to determine whether these real-world penny data refute or support the predictions you made in #3. How many of the samples contained 0, 1, or 2 heads respectively?

5. Comment on how well the Central Limit Theorem predicts the real-world results reported in #4.

Session 10

Confidence Intervals

Objectives

In this session, you will learn how to do the following:

- Construct large- and small-sample Confidence Intervals for a population mean
- Construct a large-sample Confidence Interval for a population proportion

The Concept of a Confidence Interval

A Confidence Interval is an estimate which reflects the uncertainty inherent in random sampling. To see what this means, we'll start by simulating random sampling from a hypothetical normal population, with $\mu = 500$ and $\sigma = 100$. Unlike the prior lab, though, we'll create only 20 samples, and place them into columns rather than rows.[1]

> **Calc ➤ Random Data ➤ Normal...** Generate 50 rows of data, store the results in C1–C20. Set $\mu = 500$ and $\sigma = 100$. This creates 20 samples of $n = 50$ observations.

> **Stat ➤ Basic Statistics ➤ 1-Sample Z...** This command will generate the 20 Confidence Intervals. In the dialog, select all 20 columns and specify that **Sigma** = 100.

---

[1] This is due to the way the Confidence Interval command operates; we can only construct intervals for data in columns.

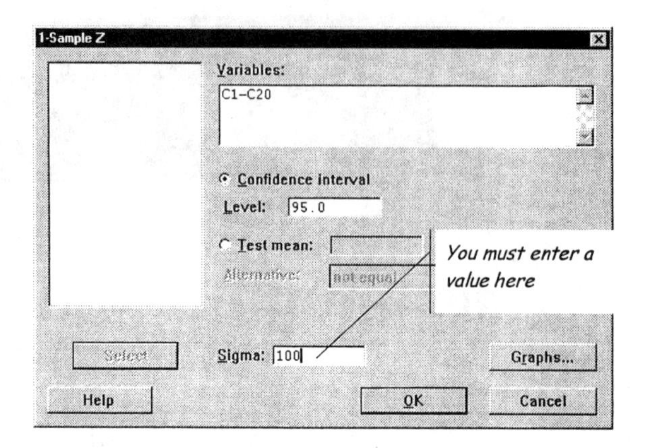

Now maximize the Session Window and look at the output; the results of one simulation are shown below.

Z Confidence Intervals

The assumed sigma = 100

Variable	N	Mean	StDev	SE Mean	95.0 % CI
C1	50	512.5	95.8	14.1	(484.7, 540.2)
C2	50	488.2	101.7	14.1	(460.5, 516.0)
C3	50	498.9	102.5	14.1	(471.2, 526.6)
C4	50	519.8	95.8	14.1	(492.0, 547.5)
C5	**50**	**535.6**	**85.6**	**14.1**	**(507.9, 563.3)**
C6	50	498.6	108.0	14.1	(470.9, 526.3)
C7	50	514.6	103.6	14.1	(486.8, 542.3)
C8	50	507.1	102.3	14.1	(479.4, 534.8)
C9	50	484.0	96.4	14.1	(456.3, 511.7)
C10	50	502.9	97.5	14.1	(475.2, 530.7)
C11	50	502.1	102.2	14.1	(474.4, 529.8)
C12	50	504.5	92.3	14.1	(476.8, 532.2)
C13	50	490.1	118.9	14.1	(462.4, 517.8)
C14	50	483.5	107.1	14.1	(455.7, 511.2)
C15	50	490.6	84.9	14.1	(462.9, 518.3)
C16	50	496.8	100.9	14.1	(469.0, 524.5)
C17	50	490.9	120.2	14.1	(463.2, 518.6)
C18	50	519.5	89.7	14.1	(491.8, 547.2)
C19	50	517.4	94.0	14.1	(489.7, 545.2)
C20	50	507.5	123.2	14.1	(479.8, 535.2)

For each of the 20 columns, there is one line of output, containing the column number, variable name, sample size, mean, standard deviation, standard error, and a 95% Confidence Interval.

> Remember that in a simulation, each of us will generate 20 different samples, and have 20 different confidence intervals. In 95% interval estimation, about 5% (1 in 20) of all possible intervals don't include μ. Therefore, you may have 20 "good" intervals, or 19, or 18 or so on.

In the sample output, one row is highlighted. In my simulation, the interval in this row lies entirely to the right of 500. Since this is a simulation, we know the true population mean (μ = 500). Therefore, the Confidence Intervals ought to be in the neighborhood of 500. ***Do all of the intervals on your screen include 500? If some do not, how many don't?***

Recall what you know about Confidence Intervals. When we refer to a 95% Confidence Interval we are saying that 95% of all possible random samples from a population would lead to an interval containing μ. Here you have generated merely 20 samples of the infinite number possible, but the pattern should become clear. Repeat this portion of the lab with 50 samples and look at the resulting intervals.

Effect of Confidence Coefficient

An important element of a confidence interval is the *Confidence Coefficient,* reflecting our degree of certainty about the estimate. By default, Minitab sets the Confidence Interval Level at 95%, but we can change that value. Generally, these coefficients are conventionally set at levels of 90%, 95%, 98%, or 99%. Let's focus on the impact of the confidence coefficient by re-constructing a series of intervals for the first sample column.

- **Edit ➤ Edit Last Dialog** Re-do the intervals for all 20 columns, using a confidence level of 90%. ***Now how many intervals include 500? How do the 90% intervals compare to the 95% intervals?***

- Do the same twice more, with confidence levels of 98% and 99%.

How do the intervals compare to one another? What is the difference from one interval to the next?

Large Samples from a Non-normal (Known) Population

Recall Session 9. We generated some large samples from a uniformly distributed population with a minimum value of 0 and a maximum of 100. In that session (see page 89), we computed that such a population has a mean of 50 and a standard deviation of 28.8675.

According to the Central Limit Theorem, the means of samples drawn from such a population will approach a normal distribution with a mean of 50 and a standard error of $28.8675/\sqrt{n}$ as n grows large. For most practical purposes, when n exceeds 30 the distribution is approximately normal; with a sample size of 50 we should be comfortably in the "large" range. As we just did, we will simulate 20 random samples, placing the observations into columns 1 through 20.

Calc ➤ Random Data ➤ Uniform Generate `50` rows of data in `C1 - C20`. **Lower endpoint** = `0`, **Upper endpoint** = `100`.

Stat ➤ Basic Statistics ➤ 1-sample Z... Select all 20 columns, set the **Confidence Interval level** = `95%`, and specify that you want **Sigma** = `28.8675`.

Again, review the output looking for any intervals that exclude 50. ***Do we still have about 95% success?***

Dealing with Real Data

Perhaps you now have a clearer understanding of a confidence interval. It is time to leave simulations behind us, and enter the realm of real data where we frequently don't know σ. For large samples (usually meaning $n > 30$), the traditional approach is to invoke the Central Limit Theorem, to estimate σ using the sample standard deviation (s), and to construct an interval using the normal distribution. In short, we find the sample standard deviation to estimate σ, and proceed as we just have. With Minitab, it is better practice to use the **1-Sample t...** command instead; it automatically uses the sample standard deviation and builds an interval using the values of the *t-distribution* rather than the normal.

Even with large samples, we should use the normal curve only when σ is known. Otherwise, the t-distribution is appropriate. In practice, the values of the normal and t-distributions become very close when n exceeds 30. With small samples, though, we face different challenges.

Small Samples from a Normal Population

If a population cannot be assumed normal, we must use large samples or non-parametric techniques such as those presented in Session 20. However, if we can assume that the parent population is normal, then small samples can be handled using the t-distribution. In Minitab, the procedural difference is quite minor, and involves using the **1-Sample t...** rather than the **1-Sample z...** on the **Stat** menu. Let's take a small sample from a population which happens to be normal: SAT scores of incoming college freshmen.

File ➤ Open Worksheet Select **Colleges**.

We will focus on the Combined SAT scores of students at public colleges, excluding private schools in case there are important differences in admission standards.

Manip ➤ Subset Worksheet Click **Condition**, and specify that `PubPvt = 1`. Clicking **OK** in both dialogs shown below will create a new worksheet containing all the variables for only Public Colleges and Universities.

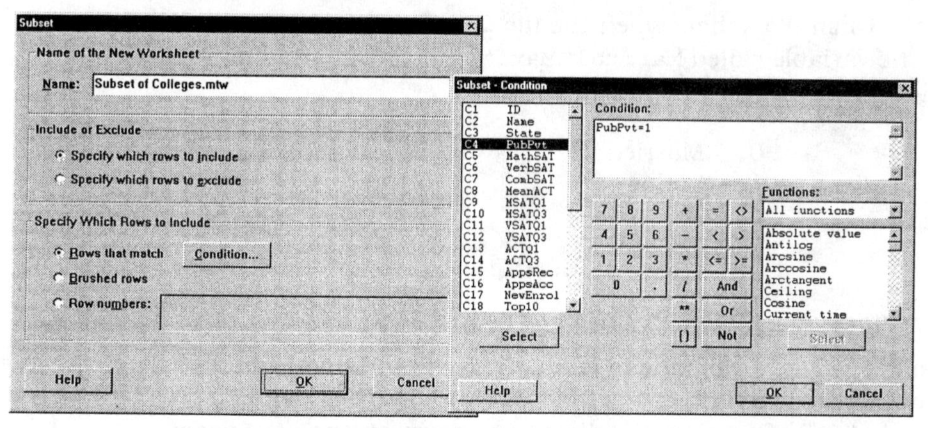

Stat ➤ Basic Statistics ➤ Display Descriptive Statistics Select `CombSAT`, and click **Graphs**. Select **Graphical Summary**.

Look at the Histogram and the Normal Curve which has been superimposed. This strongly suggests that the underlying variable is normally distributed. From this output, we can also find the mean and standard deviation, along with other statistics.

To illustrate how we'd treat a small sample from a normal population, let's treat this column as a population, and select a small random sample from it. The mean of the `CombSAT` column, for purposes of this illustration, is μ.

Calc ➤ Random Data ➤ Sample from Columns Sample 10 rows from `CombSAT`; store the sample in a new variable, **`Small`**.

Stat ➤ Basic Statistics ➤ 1-Sample t... Select `Small`, and look at the resulting interval. ***Does it contain the actual value of μ? Will everyone in the class agree with you? Explain.***

Confidence Interval for a Population Proportion[2]

The prior examples have involved quantitative variables and our goal has been to estimate a population mean. When the variable in question is qualitative, we can also use Minitab to estimate a population proportion. For example, suppose we want to estimate the proportion of Massachusetts residents who were widows or widowers in 1990.

File ➤ Open Worksheet Open the file **Census90**. Our focus is on the variable called `Marital`. `Marital` has five possible values in the file:

0	Married
1	Widowed
2	Divorced
3	Separated
4	Never married

We'll create a variable which classifies people as widowed or not.

Manip ➤ Code ➤ Numeric to text... The new variable should equal "No" when `Marital` is 0, 2, 3, or 4, and "Yes" when `Marital` is 1. Complete the dialog as shown on the next page:

---

[2] There is no equivalent command in earlier versions of Minitab, but supporting computations are easily done. See Appendix D.

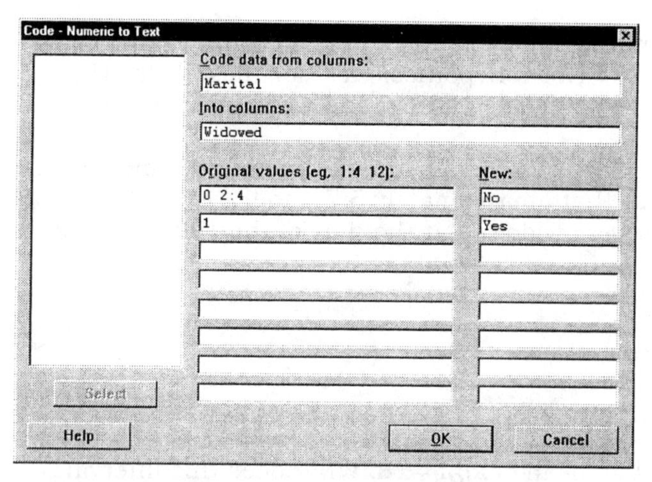

Stat ➤ Basic Statistics ➤ 1 Proportion... Just select the variable **`Widowed`**, and click **OK**.

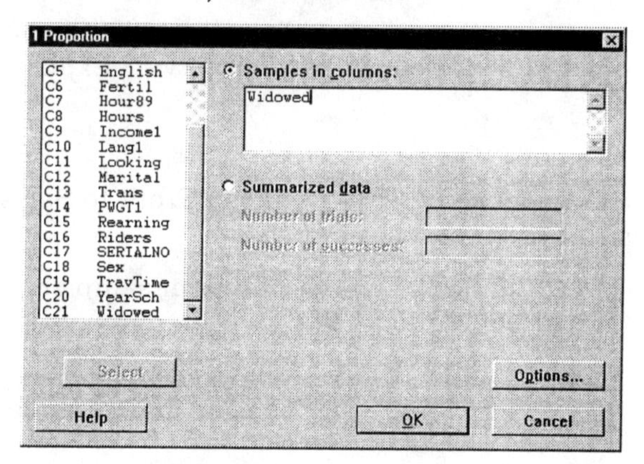

The interval estimate will appear in the Session Window:

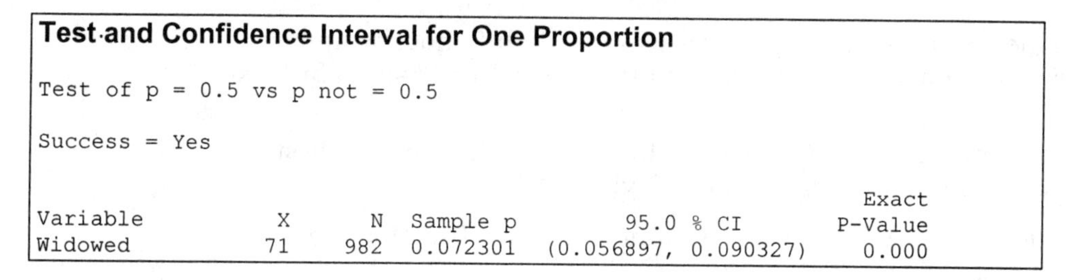

Test and Confidence Interval for One Proportion

```
Test of p = 0.5 vs p not = 0.5

Success = Yes

                                                                Exact
Variable        X      N  Sample p        95.0 % CI           P-Value
Widowed        71    982  0.072301  (0.056897, 0.090327)        0.000
```

Note that this command does double duty. It computes a confidence interval and performs a hypothesis test for these data. The next lab session deals with statistical testing, and we'll return to this example at that time.

Here the 95% Confidence Interval is from .057 to .090, indicating that we are 95% confident that approximately 5.7% to 9% of Massachusetts residents were widowed at the time of the 1990 Census.

Moving On...

Colleges

1. With the full dataset, construct a 95% Confidence Interval estimate for the mean of `RmBoard`. What does this interval suggest about what schools charge students for room and board?

2. Does this interval indicate that 95% of all students in America have room and board costs in this interval? Explain.

F500

3. Construct a 95% Confidence Interval for mean Profit to Sales Ratio. What does this interval tell us?

4. Can we consider the 1996 Fortune 500 a random sample? What would the parent population be?

5. Does this variable appear to be drawn from a normal population? What evidence would you consider to determine this?

Swimmer2

This file contains the times for a team of High School swimmers in various events. Each student recorded two "heats" or trials in at least one event.

6. Construct a 90% Confidence Interval for the mean of first times in the 100-meter Free-Style.

7. Do the same for the second times in the 100-meter Free-Style.

8. Comment on the comparison of the two intervals you've just constructed. Suggest real-world reasons which might underlie the comparisons.

Eximport

This file contains monthly data about the dollar value of U.S. exports and imports for the years 1948–1996. Consult Appendix A for variable identifications.

9. Estimate the mean value of exports to the US, excluding military aid shipments. Use a confidence level of 95%.
10. Estimate the mean value of General Imports, also using a 95% confidence level.
11. On average, would you say that the U.S. tends to import more than it exports (excluding military aid shipments)? Explain, referring to your answers to #9 and #10.
12. Estimate the mean value of imported automobiles and parts for the period covered in this file, again using a 95% confidence level.

Student

13. Construct 95% Confidence Interval estimates for the proportion of these students who:
    - Own a dog
    - Own a car
    - Know someone who has been struck by lightning
14. Do any of these intervals overlap? What does that indicate?

Session 11

One-Sample Hypothesis Tests

Objectives

In this session, you will learn how to do the following:

- Perform hypothesis tests concerning a population mean
- Perform hypothesis tests concerning a population proportion

The Logic of Hypothesis Testing

In the previous lab, the central questions involved estimating a population parameter—that is, questions like What is the value of μ? In many instances, we are less concerned with finding an estimate of a parameter as we are with comparing it to a particular value—that is, questions like Is μ more than 7? This session investigates questions of this type. To underscore the distinction between these two kinds of questions, consider an analogy from the justice system. When a crime has been committed, the question police ask is Who did this? Once a suspect has been arrested and brought to trial, the question for the jury is Did the defendant do this? Although the questions are clearly related, they are different, and the methods for interpreting evidence are also different.

Random samples provide evidence about a population. In a hypothesis test, we generally have an initial presumption about the population, much like having a defendant in court. The methods of hypothesis testing are designed to take a cautious approach to the weighing of such evidence. The tests are set up to give substantial advantage to the initial belief, and only if the sample data are very compelling do we abandon our initial position. In short, the methods of hypothesis testing provide a working definition of "compelling evidence."

In any test, we start with a *null hypothesis,* which is a statement concerning the value of a population parameter. We could, for example, express a null hypothesis as follows: "At least 75% of news coverage is positive in tone," or "The mean weekly grocery bill for a family in our city is $150." In either case, the null hypothesis states a presumed value of the parameter. The purpose of the test is to decide whether data from a particular sample are so far at odds with that null hypothesis as to force us to reject it in favor of an *alternative* hypothesis.

An Artificial Example

We start with some tests concerning a population mean, and return to the first simulation we conducted in introducing confidence intervals. In that case, we simulated drawing twenty random samples from a normal population with $\mu = 500$ and $\sigma = 100$. We'll do the same thing again, understanding that there are an infinite number of possible random samples, each with its own sample mean. Though it is likely that our samples will each have a sample mean of about 500, it is *possible* that we will obtain a sample with a mean so far from 500 that we might be convinced that the population mean is not 500.

What's the point of the simulation? Remember that this is an *artificial* example. Usually, we **do not know** the truth, and are trying to infer it from a random sample. Ordinarily, we would not know the truth about μ; we would have one sample, and we would be asking if this sample is consistent with the hypothesis that $\mu = 500$. This simulation can give us a feel for the risk of an incorrect inference based on any single sample.

Calc ➤ Random Data ➤ Normal Generate 50 rows of data, and store the results in C1–C20. Set $\mu = 500$ and $\sigma = 100$, as you did in the previous session.

Stat ➤ Basic Statistics ➤ 1-Sample Z... The command can either generate intervals or perform a test based on the normal distribution. Select all 20 columns, and specify that you want to **Test** (a) **mean** against a hypothetical value of 500, assuming a population sigma of 100. Note that you don't specify a significance level.

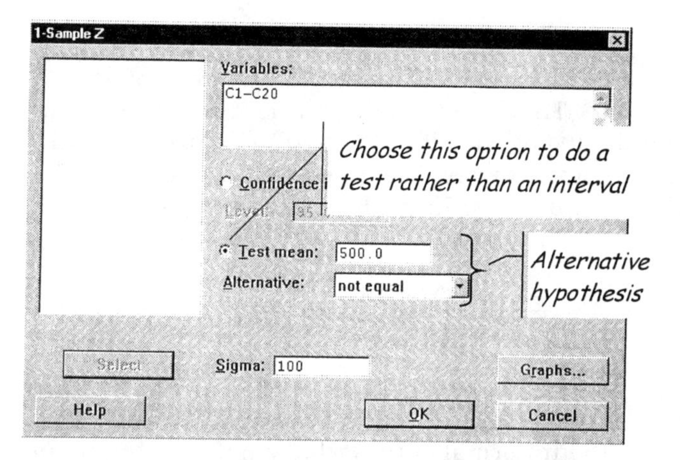

Now look at Session Window (sample shown here). Remember that your output will be different since this is a simulation.

Z-Test

Your hypotheses

```
Test of mu = 500.0 vs mu not = 500.0
The assumed sigma = 100
```

Variable	N	Mean	StDev	SE Mean	Z	P
C1	50	496.6	98.6	14.1	-0.24	0.81
C2	50	485.7	100.2	14.1	-1.01	0.31
C3	50	502.1	98.5	14.1	0.15	0.88
C4	50	507.5	88.1	14.1	0.53	0.59
C5	50	498.3	110.9	14.1	-0.12	0.90
C6	50	488.5	97.8	14.1	-0.81	0.42
C7	50	526.0	103.0	14.1	1.84	0.066
C8	50	492.0	123.1	14.1	-0.57	0.57
C9	50	506.9	97.2	14.1	0.49	0.62
C10	50	479.7	105.2	14.1	-1.44	0.15
C11	50	496.5	94.2	14.1	-0.24	0.81
C12	50	482.7	94.4	14.1	-1.23	0.22
C13	**50**	**536.7**	**88.9**	**14.1**	**2.60**	**0.0095**
C14	50	495.3	88.1	14.1	-0.33	0.74
C15	50	479.9	100.8	14.1	-1.42	0.15
C16	50	516.6	95.5	14.1	1.17	0.24
C17	50	475.9	113.6	14.1	-1.70	0.089
C18	50	474.0	98.4	14.1	-1.84	0.066
C19	50	502.2	112.9	14.1	0.16	0.88
C20	50	514.7	104.7	14.1	1.04	0.30

The output reports the null and alternative hypotheses, and summarizes the results of these random samples. In this example, the first sample mean was 496.6: This is below 500, but is it so far below as to cast serious doubt on the hypothesis that $\mu = 500$? The **test statistic** gives us a relative measure of the sample mean, so that we can judge how consistent it is with the null hypothesis. In a large sample test with a known population σ, the test statistic is computed as follows:

$$z = \frac{\bar{x} - \mu}{\sigma/\sqrt{n}} = \frac{496.6 - 500}{100/\sqrt{50}} = -0.24$$

This is the value reported in the Z column of the output. In other words, 496.6 is only 0.24 standard errors below the hypothesized value of μ. Given what we know about normal curves, that's not very far off at all. It is quite consistent with the kinds of random samples one would expect from a population with a mean value of 500. In fact, we could determine the likelihood of observing a sample mean more than 0.24 standard errors away from 500. That likelihood is called the "p-value" and it appears in the last column. In this instance, $p \leq 0.81$.

One way of thinking about the p-value is that if you were to reject the null hypothesis on the basis of this test, there is a probability of at most 0.81 that you are making a Type I error.[1] Since that probability is so high, you would be well advised against rejecting the null hypothesis in this instance.

Look down the list of z-scores and p-values in the output on the prior page. Note the line which is highlighted. In this particular random sample, the z-score was 2.60, and the p-value was only 0.0095. For this sample, at a significance level of $\alpha = .05$, we would *reject* the null hypothesis, and erroneously conclude that the population mean is not equal to 500.

Since this is a simulation, we know that the true population mean *is* 500. Consequently, we know that the null hypothesis really is true, and that most samples would reflect that fact. We also know that random sampling involves uncertainty, and that the population does have variation within it. Therefore, some samples will have sufficiently small p-values that we would actually reject the null hypothesis. In this simulation, one sample in 20 (5%) leads to this erroneous conclusion.

---

[1] Consult your textbook for further information about Type I errors and about p-values.

What happened in your simulation? Assuming a desired significance level of α = .05, would you reject the null hypothesis based on any of these samples? What kinds of results do you think other people in the class generated?

A More Realistic Case: We Don't Know Sigma

Simulations are instructive, but are obviously artificial. This simulation is unrealistic in at least two respects—in real studies, we generally don't know μ or σ, and we have only one sample to work with. Let's see what happens in a more realistic case.

File ➤ Open Worksheet... Bodyfat

Each of us carries around different amounts of body fat. There is considerable evidence that important health consequences relate to the percentage of fat in one's total body mass. According to one popular health and diet author, fat constitutes 23% of total body mass (on average) of American adult males.[2]

Our data file contains body fat percentages for a sample of 252 males. Is this sample is consistent with the assertion that the mean bodyfat percentage of the American adult male population is 23%?

Since we have no reason to suspect otherwise, we can assume that the sample does come from a population whose mean is 23%, and establish the following null and alternative hypotheses for our test:

H_o: $\mu = 23$
H_A: $\mu \neq 23$

The null hypothesis is that this sample comes from a population whose mean is 23 percent body fat. The two-sided alternative is that the sample was drawn from a population whose mean is other than 23.

This test has little practical importance, so we can tolerate a fairly high *significance level* (α), such as 0.10. In other words, if we reject H_o, we'll settle for being 90% confident in our conclusion. Thus, the decision rule is this: We will reject the null hypothesis only if our test statistic has an associated p-value of 0.10 or less.

The dataset represents a large sample (n = 252), so we can rely on the Central Limit Theorem to assert that the sampling distribution is approximately normal (assuming a random sample). Unfortunately, we don't know the population standard deviation, σ, which we need for the

[2] Sears, Barry. *The Zone.* (New York: HarperCollins, 1995).

z-procedure. In Minitab, the most direct and expeditious approach is to use the one-sample t-test.

Stat ➤ Basic Statistics ➤ 1-Sample t... Select **FatPerc**, enter the hypothetical mean value of 23, and use the default setting of "**not equal**" as the alternative hypothesis.

Now look in the Session Window at the "T-Test" results.

T-Test of the Mean

```
Test of mu = 23.000 vs mu not = 23.000

Variable       N      Mean     StDev   SE Mean        T          P
FatPerc      252    19.151     8.369     0.527    -7.30     0.0000
```

Here the value of the test statistic, t, is a whopping –7.3 standard errors. That is to say, the sample mean of 19.15% is extremely far from the hypothesized value of 23%. The p-value of approximately 0.0 suggests that we should confidently *reject* null hypothesis, and conclude that these men are selected from a population with a mean body fat percentage of something other than 23.

A Small-Sample Example

In the previous example, we used a t-test with a large sample. What happens when the sample is small? You may have learned that any $n > 30$ is "large," and that the Central Limit Theorem can apply in such cases. While that is a good rule of thumb, the t-test is preferable for samples in the neighborhood of 30 observations, especially since σ is generally unknown in real-world data.

One bit of 'conventional wisdom' is that it is unhealthy to have a total cholesterol level in excess of 200. We have a data file with sample observations of 30 people who had recent heart attacks. One might suspect that their cholesterol levels would be above that level. Specifically, we'll consider their levels measured two weeks after the heart attack, using a variable called '14-day.' Let $\mu_{14\text{-day}}$ represent the mean of '14-day.'

Formally, our hypotheses will be:

$$H_o: \mu_{14\text{-day}} \leq 200$$
$$H_A: \mu_{14\text{-day}} > 200$$

File ➤ Open Worksheet... Open **Cholest**

Because this is a small-sample, we may apply the t-distribution to a hypothesis test *only* if the underlying population is normally distributed (or at least bell-shaped and symmetrical). We can check the assumption of a normal population by looking at the histogram of the sample data. If it is reasonably bell-shaped, we can proceed.

Stat ➤ Basic Statistics ➤ Display Descriptive Statistics Select the variable `14-Day`, click **Graphs**, and specify that you want the **Graphical Summary**.

Descriptive Statistics

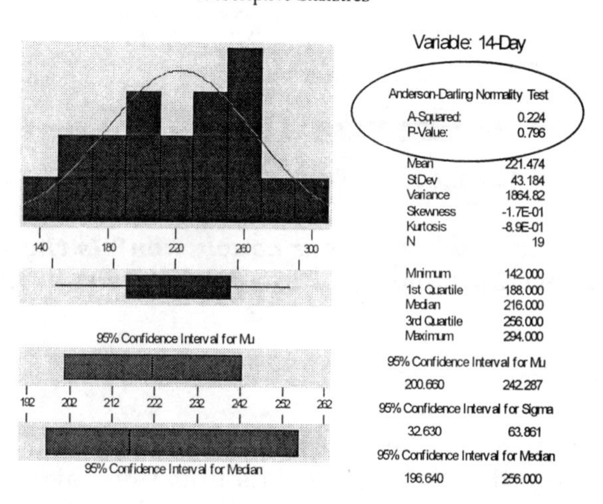

This command computes all of the usual descriptive statistics for the variable, displays a histogram of the data with a normal curve superimposed on the histogram, and also performs a statistical test of the null hypothesis that the sample was drawn from a normal population.[3] On the output, there is a reference to the *Anderson-Darling Normality Test*, with a test statistic called A-squared, and an associated p-value. When the p-value is very small (under .05 or so), we have enough evidence to reject the normality assumption. In this case, the p-value is quite high, suggesting that we may proceed with our t-test.

---

[3] This is one of a class of tests known as "Goodness of Fit" tests. Some of these procedures are presented more fully in Session 13.

Stat ➤ Basic Statistics ➤ 1-Sample t... In the dialog, select the variable `14-day`, specify a hypothesized value of 200, and an alternative of greater than.

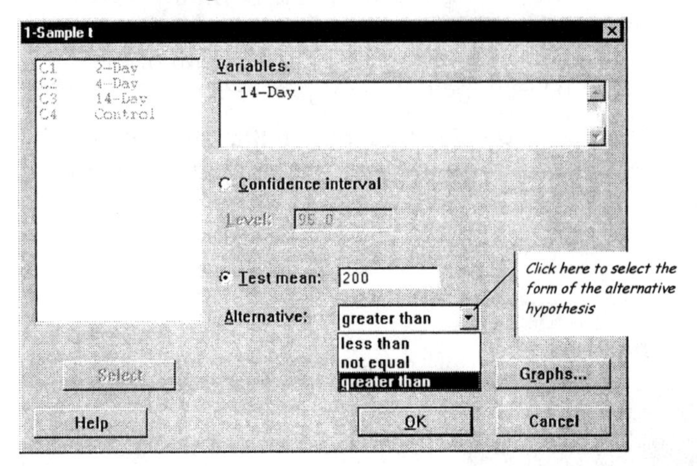

On the basis of this test, what is your conclusion? Is the mean cholesterol level above 200 among people with recent heart attacks?

A Test Involving a Proportion[4]

Let's return to our data about the marital status of Massachusetts residents. According to national data from1993, about 8.9 percent of Americans aged 16 and older were divorced. Let's see if the Massachusetts sample is consistent with the national proportion. Our null hypothesis will be that in Massachusetts, the proportion of divorced adults is .089; we'll test it against the two-sided alternative.

Recall that we needed to manipulate the data a bit to select the adults of marriage age. We'll follow the same preliminary procedure here in order to create a new binary variable which classifies people as divorced or not divorced. After opening the **Census90** worksheet, do this:

Manip ➤ Subset Worksheet... Create a subset of the entire Census file, selecting those observations which satisfy the **Condition** that `Age > 15`.

---

[4] There is no equivalent command in earlier versions of Minitab. See Appendix D.

Manip ➤ Code ➤ Numeric to text... The new variable (call it `Divorced`) should equal "No" when `Marital` is 0, 1, 3, or 4, and "Yes" when `Marital` is 2 (refer to Session 10, page 99, for a similar illustration of coding).

Stat ➤ Basic Statistics ➤ 1 Proportion... Select `Divorced`, and click **Options**. In the **Options** dialog, specify a hypothetical value of `0.089`, and a **not equal** to alternative.

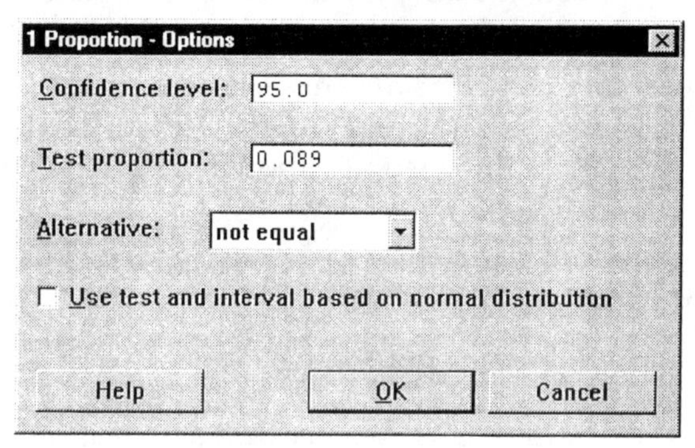

The output will look like this:

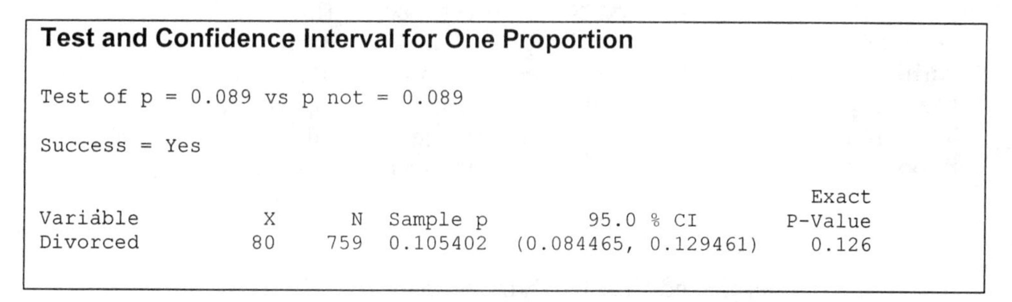

Test and Confidence Interval for One Proportion

Test of p = 0.089 vs p not = 0.089

Success = Yes

Variable	X	N	Sample p	95.0 % CI	Exact P-Value
Divorced	80	759	0.105402	(0.084465, 0.129461)	0.126

Here, the sample proportion is about 10% (0.1054). We don't see a test statistic reported, but we do get a confidence interval and p-value. The noteworthy point is that the p-value is reasonably large ($p > .10$), suggesting that we fail to reject the null hypothesis. In other words, although the divorce rate in this sample exceeds 10%, that is not so far above 8.9% as to cast doubt on the null hypothesis. Also note that the 95% confidence interval contains 8.9%, indicating that the sampling error is sufficiently large as to include the national proportion in an interval estimate.

Moving On...

Now apply what you have learned in this session. You can use 1-sample t- or proportion tests for each of these questions. Explain what you conclude from each test, and why.

Bev

1. Test the null hypothesis that the mean Current Ratio for this entire sample of firms is less than or equal to 3.0.
2. Using the **Subset** command, isolate the bottled and soft drink firms (SIC = 2086) and their current ratios. Use an appropriate test to see if the current ratio for these companies is *significantly different* from 3.0.[5]
3. What about the malt beverage firms (SIC = 2082)? Is their ratio significantly different from 3.0?

Paworld

This dataset contains multiple yearly observations for a large sample of countries around the globe.

4. One of the variables is called "C," and represents the percentage of Gross Domestic Product (GDP) consumed within the country for the given year. The mean value of C in the sample is 64.93%. Was this average value *significantly* less than 65% of GDP? (i.e., Would you reject a null hypothesis that $\mu \geq 65.0$ at a significance level of $\alpha = .05$?)

Student

5. It is generally estimated that left-handed people make up about 10% of the US population. Do these students appear to be drawn from such a population?
6. Several years back, about 30% of my students reported having smoked at least one cigarette in the past month. Based on the evidence in this sample, has that proportion increased recently?

---

[5] If you are using a Minitab version prior to Release 12, there is no Subset command. Consult Appendix D for an alternate approach.

Session 12

Two-Sample Hypothesis Tests

Objectives

In this session, you will learn to do the following:

- Perform hypothesis tests concerning the difference in means of two populations
- Perform hypothesis tests concerning the difference in means of two "matched" samples drawn from a population
- Perform hypothesis tests concerning the difference between two population proportions

Working with Two Samples

In the prior lab session, we learned to make inferences about the mean of a population. Often our interest is in *comparing the means of two distinct populations.* To make such comparisons, we must select two independent samples, one from each population.

For samples to be considered independent, we must have no reason to believe that the observed values in one sample could affect or be affected by the observations in the other, or that the two sets of observations arise from some shared factor or influence.

We know enough about random sampling to predict that any two samples will likely have different sample means *even if they were drawn from the same population.* We anticipate some variation between any two sample means. Therefore, the key question in comparisons of samples from two populations is this: Is the observed difference between two sample means large enough to convince us that the populations have different means?

Sometimes, our analysis focuses on two distinct groups within a single population, such as females and males in one student body. Our first example does just that. For starters, let's test the radical theory that male colleges students are taller than female college students. Open the **Student** worksheet.

We can restate our theory in formal terms as follows:

$$H_o: \mu_f - \mu_m \geq 0$$
$$H_A: \mu_f - \mu_m < 0$$

Note that the hypothesis is expressed in terms of the *difference* between the means of the two groups. The null says that men are no taller than women, and the alternative is that men are taller on average.

In an earlier lab, we created histograms to compare the distribution of heights for these male and female students. At that time, we visually interpreted the graphs. Now let's do the test, and generate another graph to help illuminate the comparison.

You may have learned that a 2-sample t-test requires three conditions:

- Independent samples
- Normal populations
- Equal population variances (for small samples)

The last item is not actually required to perform a t-test. The computation of a test statistic is different when the variances are equal, but Minitab can readily handle situations where we do not assume that $\sigma_1^2 \neq \sigma_2^2$. When variances are in fact unequal, treating them as equal may lead to a seriously flawed result. You'll see that the distinction is minor in terms of using the software.

As for the assumption of normality, the t-test is reliable so long as the samples suggest symmetric, bell-shaped data without gross departures from a normal distribution. Since human height is generally normal for each sex, we should be safe here. Nevertheless, we are well-advised to examine our data for the bell shape.

Stat ➤ Basic Statistics ➤ Display descriptive statistics... We want to describe `Ht`, grouping **by variable** `Gender`, and generate **Histograms of data, with normal curve** superimposed for both groups. By now, you should know your way around this dialog. Tiling the two graphs shows this:

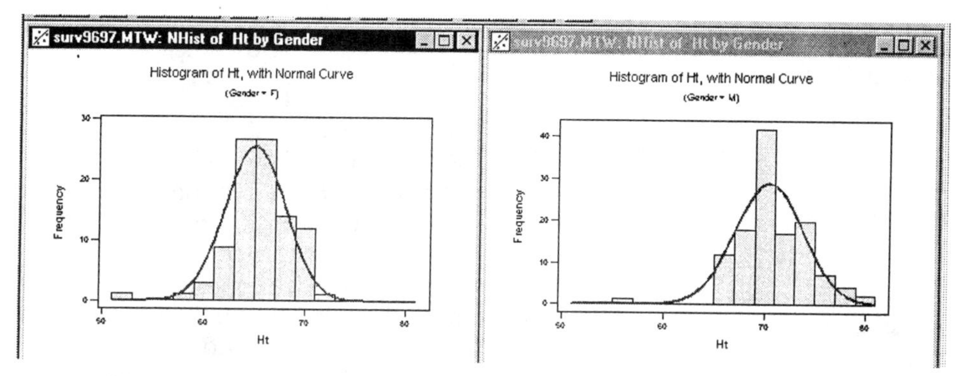

Though not perfectly normal, these are reasonably symmetrical and bell-shaped, and suitable for performing the t-test.[1]

Stat ➤ Basic Statistics ➤ 2-Sample t... Since all of the height data is in one column, we choose **Samples in one column**, select `Ht` for **Samples** and `Gender` for **Subscripts.**[2] Also choose a **less than** alternative. Our gender variable is coded F and M; Minitab will treat the 'F' observations as the first sample, since F precedes M alphabetically.

Before clicking **OK,** click on the **Graphs** button, and choose **Boxplots.** Click **OK** in the **2-Sample t-Graphs** dialog.

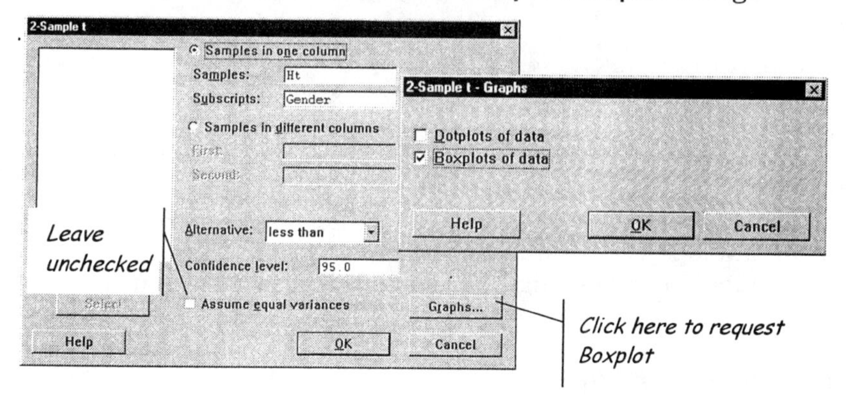

---

[1] Session 20, on nonparametric techniques, addresses the situation in which we have non-normal data.

[2] Minitab refers to this data arranged this way as "stacked." In most of our examples, we'll use stacked data. See Appendix C for a full discussion of the reasons for stacking or unstacking data.

Leave the **Assume equal variances** box unchecked, and click **OK**. You'll see this graph:

In the boxplot, it appears that most of the women are indeed shorter than most of the men. The red dots near the median line in each box represent the location of the sample means. The test results in the Session Window will tell us whether the difference in sample means is so large as to reject the null hypothesis in favor of the alternative.

Two Sample T-Test and Confidence Interval

```
Two sample T for Ht

Gender          N       Mean      StDev   SE Mean
F              95      65.03       2.98      0.31
M             123      70.45       3.38      0.30

95% CI for mu (F) - mu (M): ( -6.27,  -4.56)
T-Test mu (F) = mu (M) (vs <): T = -12.54  P = 0.0000  DF = 212

* NOTE * N missing = 1
```

We interpret the output much in the same way as in the one-sample test. The test statistic t equals –12.54. Since the p-value here is approximately 0, we would reject the null hypothesis in favor of the alternative, and conclude the mean height for females is less than that for males.

Matched vs. Independent Samples

In the prior examples, we have focused on differences inferred from two independent samples. Sometimes, though, our concern is with the *change* in a single variable observed at two points in time. For example, to evaluate the effectiveness of a weight-loss clinic with 50 clients, we need to assess the change experienced by each individual, and not merely the collective gain or loss of the whole group.

We could regard such a situation as an instance involving two different samples. These are sometimes called “matched samples,” “repeated measures,” or “paired observations.” [3] Since the subjects in the samples are the same, we pair the observations for each subject in the sample, and focus on the difference between two successive observations or measurements. We'll see how it works by investigating per capita income around the world.

Open the worksheet called **World90**.

This file contains several variables describing demographic and economic status of 42 countries. Two of the variables measure real per capita GDP in 1988 and 1990.[4] Specifically, RGDPCH and RGDP88 represent the per capita GDP values for the years 1990 and 1988. Suppose we're interested in knowing if per capita GDP increased significantly in the two year period.

We are performing this test to see if there is evidence to suggest that GDP increased; that is the suspicion which led us to perform the test. Therefore, in this test, our null hypothesis is to the contrary, which is that there has been no change. Let μ equal the mean change in per capita GDP from 1988 to 1990.

H_0: $\mu \leq 0$ {GDP did not increase}
H_A: $\mu > 0$ {GDP did increase}

In most countries, we expect the 1990 figure to be similar to that nation's 1988 figure. This is why we must treat the samples as dependent in consecutive years. Therefore, we use a different procedure to conduct the test.

Stat ➤ Basic Statistics ➤ Paired t... In the dialog, select RGDPCH as the first variable, and RGDP88 as the second. Also, request a **Histogram of differences** by clicking on the **Graphs** button. In the **Options** dialog, select a **greater than** alternative.

---

[3] Matched samples are not restricted to “before and after” studies. Your text will provide other instances of their use. The goal here is merely to illustrate the technique in one common setting.

[4] GDP stands for Gross Domestic Product, which is the aggregate value of all goods and services produced within a country. “Real” GDP represents the value, net of inflation.

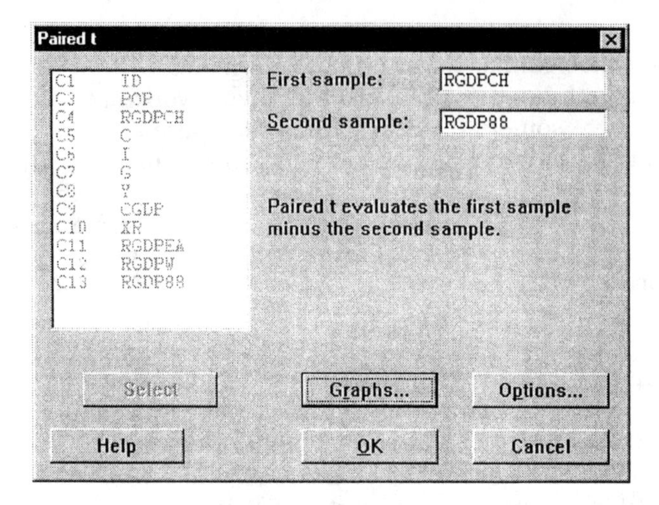

The pairwise differences are shown here, and the test results from the Session Window below them.

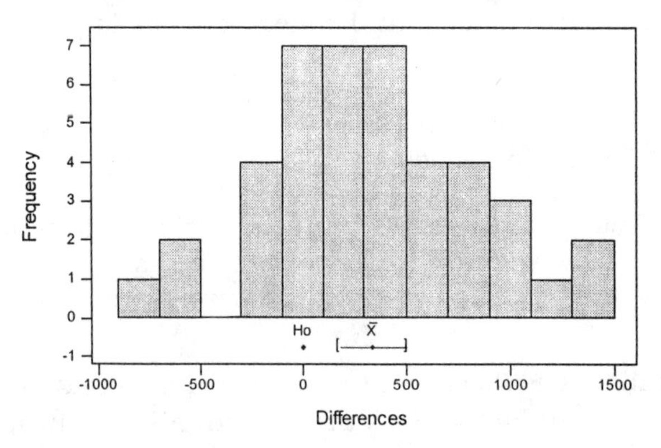

Paired T-Test and Confidence Interval

```
Paired T for RGDPCH - RGDP88

                  N      Mean     StDev   SE Mean
RGDPCH           42      7907      5410       835
RGDP88           42      7569      5230       807
Difference       42     338.0     510.5      78.8

95% CI for mean difference: (178.9, 497.1)
T-Test of mean difference = 0 (vs > 0): T-Value = 4.29 P-Value = 0.000
```

What do you conclude? Was the observed increase statistically significant? In the histogram, what does the point marked H_0 represent? What does the red line marked $\bar{x}$ represent?

As noted earlier, we correctly treat this as a paired sample t-test. But what would happen if we were to (mistakenly) treat the data as two *independent* samples? Let's try it and see.

Stat ➤ Basic Statistics ➤ 2-sample t... Now, we have data in two columns (`RGDPCH,RGDP88`), and our alternative hypothesis is that the first sample is **greater than** the second (i.e., GDP in 1990 was greater than it was in 1988).

How does this result compare to prior one? In the correct version of this test, we saw a large, significant increase—the t-statistic was in excess of 5. ***What happens in this test? Why the difference? So did GDP increase in from 1988 to 1990 or not?***

This example illustrates the importance of knowing which test applies in a particular case. Minitab (or any software package) is readily able to perform the computations either way, but the onus is on the analyst to know which method is the appropriate one. As you can see, getting it "right" makes a difference—a large effect "disappears" when viewed through the lens of an inappropriate test.

Comparing Two Proportions

Suppose we want to know whether male college students are more likely than female students to own a car. After all, one common image of the American male is that he is particularly devoted to his automobile. Let's return to the Student data to see if it can shed light on the question. Click in the **Student** Data Window to select it, or re-open the worksheet if you previously closed it.

This question involves a binary categorical variable, car ownership (`Owncar`) and we want to compare the proportion of owners across `Gender`. These are stacked data, with all students' responses to the car ownership question in one column, and subscripts identifying gender in another.

Stat ➤ Basic Statistics ➤ 2 Proportions... As shown in the completed dialog box, `OwnCar` is the **Sample** and `Gender` contains the **Subscripts**. Click **Options** to specify a **less than** alternative hypothesis.

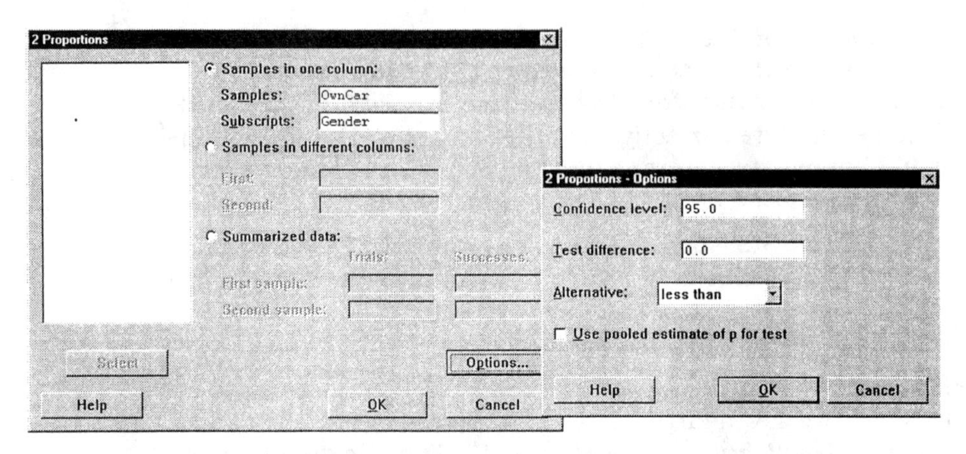

Before looking at the results, there is one restriction in the use of this command: The sample may not contain missing data, or the command will not execute. If there are missing observations, subset the worksheet, omitting rows where that variable is missing data. Alternatively, use a cross-tabulation to summarize the number of 'successes' for each sample, and then simply enter the figures in the dialog under the **Summarized data** option.

The results are these:

Test and Confidence Interval for Two Proportions

```
Success = Y

Gender              X       N  Sample p
F                  60      96  0.625000
M                  81     123  0.658537

Estimate for p(F) - p(M):   -0.0335366
95% CI for p(F) - p(M):   (-0.161605, 0.0945315)
Test for p(F) - p(M) = 0 (vs < 0):   Z = -0.51  P-Value = 0.304
```

This command computes the sample proportions for each sample, along with point and interval estimates for the difference between the two population proportions. As for the test, the test statistic follows an approximate normal distribution; in this case, the p-value is 0.304, which leads us to conclude that we fail to reject the null. These samples provide no convincing evidence that male students own cars in greater proportions than females.

Moving On...

Use the techniques presented in this lab to answer the following research questions. Justify your conclusions by citing appropriate test statistics and p-values. Unless otherwise noted, use $\alpha = .05$.

Student

1. Do commuters and residents earn significantly different mean grades?
2. Do car owners have significantly fewer accidents, on average, than non-owners? (Hint: Look closely at Session output.)
3. Do dog owners have fewer siblings on average than non-owners?
4. Many students have part-time jobs while in school. Is there a significant difference in the mean number of hours of work for males and females who have such jobs? (Omit students who do not have outside hours of work.)
5. Are males more or less likely than females to be dog owners?

Colleges

6. For each of the variables listed below, ***explain why you might expect to find significant differences between means for public and private colleges.*** Then, test to see if there is a significant difference between public and private colleges.
   - `CombSAT` (Mean combined SAT scores of incoming freshmen)
   - `Top10` (Percent of incoming freshmen in the top 10% of their High School class)
   - `FTUnder` (Number of full-time undergraduate students)
   - `Tuit_In` (Tuition charges for in-state students)
   - `RmBoard` (Room and board charges)
   - `GradRate` (Percent of students who graduate within four years)

Swimmer1 *(Note: These are challenging problems.)*

7. Do individual swimmers significantly improve their performance between the first and second recorded times?

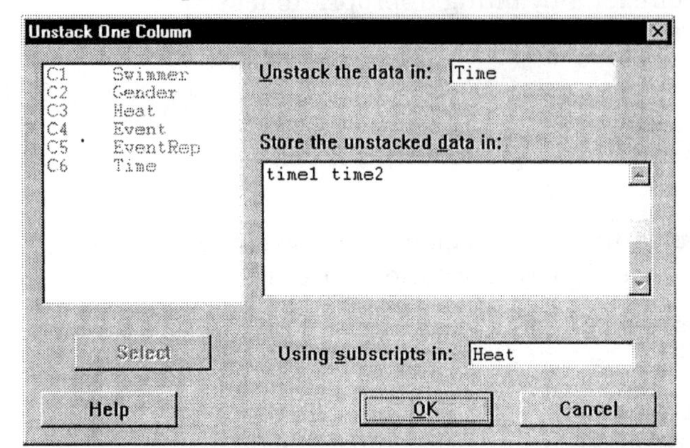

NOTE: To answer this question, you must "unstack" the Time data.

Manip ➤ Stack/Unstack ➤ Unstack One Column

Complete the dialog as shown. Do a paired-sample test using the two new columns.

8. In their second times, do swimmers who compete in the 50-meter Free-Style swim faster than those who compete in the 50-meter breaststroke? (Subset the worksheet.)

9. In the 50-meter Free-Style, do the men swim faster than the women? (Select either heat and subset the worksheet.)

Water

10. Is there statistically significant evidence here that water resources sub-regions were able to reduce irrigation conveyance losses (i.e., leaks) between 1985 and 1990?

11. Did mean per capita water use change significantly between 1985 and 1990?

Cholest

12. Comparing the heart attack patients on Day 2 to the control patients, we might expect the latter group to have lower cholesterol readings. Do the data support that conclusion at a 0.05 significance level?

13. Is there a significant reduction in cholesterol levels for the heart attack patients between Day 2 and Day 14?

Session 13

Chi-Square Tests

Objectives

In this session, you will learn to do the following:

- Perform Chi-Square Goodness of Fit tests
- Perform Chi-Square Tests of Independence

Review of Qualitative vs. Quantitative Data

All of the tests we have studied thus far have been appropriate exclusively for quantitative variables. The tests presented in this session are suited to analyzing *qualitative* or discrete quantitative variables, and the relationships between two such variables. The tests fall into two categories: Goodness of Fit tests and Tests for Independence.

Goodness of Fit Testing

When we construct a mathematical model of a process or phenomenon, we often begin with a sample, and develop a model, which reliably describes or simulates the empirical data. Then we can test the model by comparing another sample to it. For instance, we can use the binomial distribution to predict the outcomes of flipping a fair coin 50 times, and then compare sample results to the binomial model's predictions.

A different real-world activity might nearly meet the theoretical requirements of the Poisson distribution, but a single sample's relative frequencies might not perfectly match the Poisson probabilities. In each case, we might want to ask whether observed data so closely follow the theoretical distribution that we could use the theoretical distribution as a model of the activity.

Tests which help to answer such questions are called "Goodness of Fit" tests. In earlier labs, you have seen the Anderson-Darling Normality Test (page 109), which helps us to decide whether a sample has been drawn from a normal population. The tests in this lab are also Goodness of Fit tests, all of which rely on the Chi-Square distribution.

A First Example: Simple Genetics

In the 1860's, Gregor Mendel conducted a series of experiments, which formed the basis for the modern study of genetics. In several sets of experiments on peas, Mendel was interested in the heredity of one particular characteristic—the texture of the pea seed. He had observed that pea seeds are always smooth or wrinkled.

Mendel determined that smoothness is a *dominant* trait. In each generation, an individual pea plant receives genetic material from two parent plants. If either parent plant transmitted a smoothness gene, the resulting pea seeds would be smooth. Only if both parents transmitted wrinkled genes would the offspring pea be wrinkled.

This is the logical equivalent of flipping two coins, and asking about the chances of getting two heads. If the "parent" peas each have one smooth and one wrinkled gene (SW), then an offspring can have one of four possible combinations: SS, SW, WS, or WW. Since smoothness dominates, only the WW pea seed will have a wrinkled appearance. Thus, the probability of a wrinkled offspring is just 0.25.

Over a number of experiments, Mendel assembled data on 7324 second-generation hybrid pea plants. If the model just described is correct, we would expect one-fourth of the plants (1831) to be wrinkled, and the remaining 5493 to be smooth. In his trials, Mendel found 5474 smooth plants, and the rest were wrinkled.

We can use a Chi-Square Goodness of Fit test to see whether the model accurately predicts the laboratory findings. Our null hypothesis is that the peas *do* follow Mendel's prediction:

H_o: $p_{wrinkled} = .25$, $p_{smooth} = .75$
H_A: H_o is false

The Chi-Square Goodness of Fit test compares the observed sample frequencies to the *expected frequencies* that we would find in a sample of the same size, if the hypothesized percentages are accurate.

Hence, if the probability of *wrinkled* is 0.25, then we expect to find 1831 wrinkled pea plants in a sample of 7324 plants tested.

Above, we have the observed and the expected frequencies. We can use Minitab to help conduct the test.

- In the blank worksheet, label the first three columns **Texture**, **Observed**, and **Expected**.

- Enter the data, as shown in this completed worksheet below:

Worksheet 1 ***

	C1-T	C2	C3
↓	Texture	Observed	Expected
1	Smooth	5474	5493
2	Wrinkled	1850	1831
3			

In this test, the test statistic is given by this formula:

$$\chi^2 = \sum_{i=1}^{k} \frac{(o_i - e_i)^2}{e_i}$$

where

o_i is the observed frequency of the *i*th category and
e_i is the expected frequency of the *i*th category and
k is the number of categories

The test statistic follows a Chi-Square distribution with k - *1* degrees of freedom. We can use the Calculator to compute the test statistic, and the Probability Distribution function on the Calc menu to determine the significance level of the test statistic. We start by computing the test statistic, and storing it in the Minitab constant, **k1**.

> For Goodness of Fit tests, it may be simpler to use Minitab to calculate the test statistic, and then consult a Chi-Square table. In this example, we show how to do the entire test.

- **Calc ➤ Calculator** Type **K1** in the box marked **Store result in.** In the **Expression** box, copy the expression shown in the dialog on the next page.

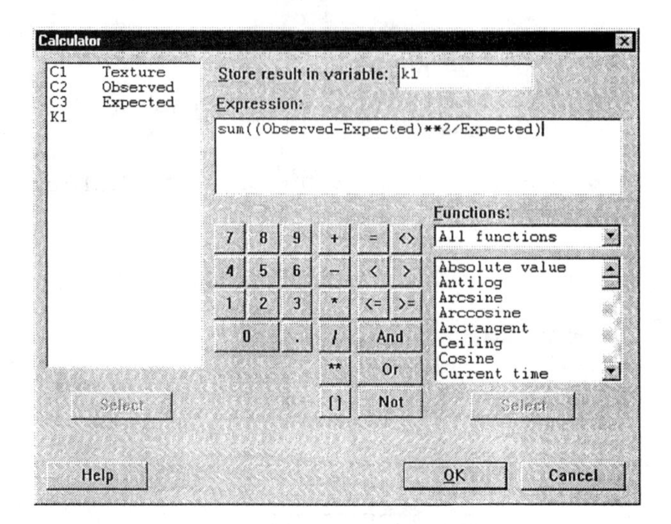

Nothing changes in the Session or Data Windows, but now the Minitab constant **K1** contains the value of the test statistic. Open the Info window (**Window ➤ Info**) to see the value of K1 (0.26288).

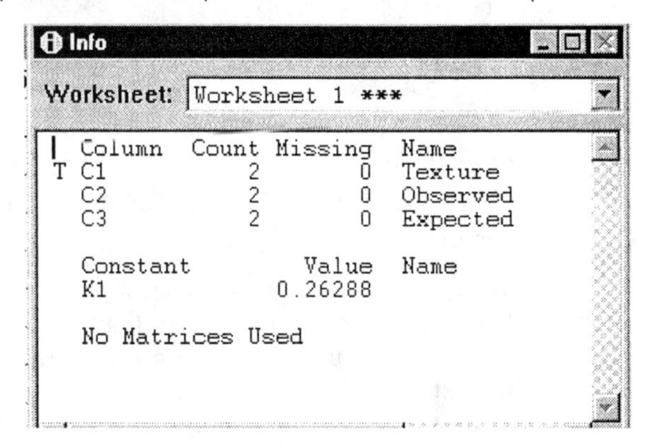

Now let's find the p-value for the statistic. The simplest approach is to find the cumulative probability between 0 and the test statistic; the p-value is the complement of this.

Calc ➤ Probability Distributions ➤ Chi-Square Choose **Cumulative probability.** We need to specify the number of degrees of freedom, which is *k - 1*, or in this case, 2 - 1 = 1. Enter 1 in **Degrees of freedom**. Choose **Input constant**, and type K1.

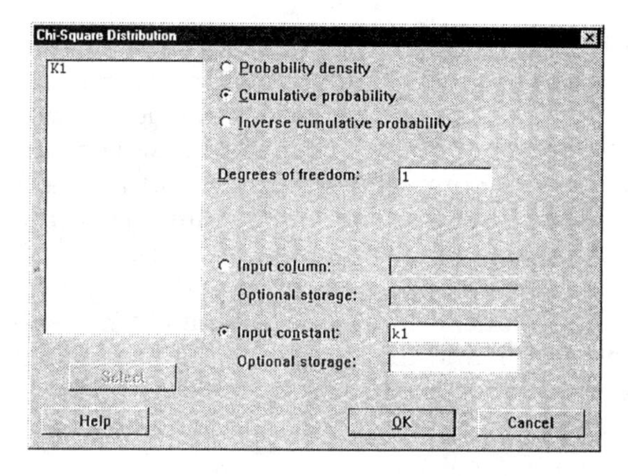

In the Session Window, we find the cumulative probability of the test statistic, which in this case is .3919. The p-value, therefore, is .6081, and we clearly fail to reject the null hypothesis. In other words, the experimental results are consistent with Mendel's model of inheritance.

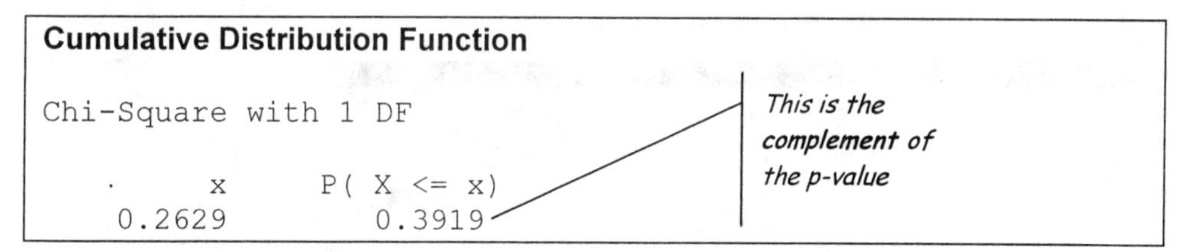

Cumulative Distribution Function

```
Chi-Square with 1 DF

        x      P( X <= x)
   0.2629          0.3919
```


Testing for Independence

The Chi-Square distribution is also useful for testing whether two qualitative or discrete variables are statistically independent of one another. Chi-Square goodness of fit tests require multiple steps in Minitab. In contrast, tests for independence are quite straightforward.

The logic of the test is much the same as in the Goodness of Fit test. We start with a theory that predicts a set of frequencies. We then compare observed frequencies to the predicted ones. If the deviations between the observed and expected are sufficiently large, we reject the initial theory as incorrect.

For instance, in the **Student** dataset (open it now), we have a binary variable which represents whether or not the student owns a car (`OwnCar`), and another representing whether the student is a resident, living on-campus, or a commuter (`Res`). It is reasonable to think that these two variables might be related, with commuters having a greater need for an automobile. We can use a Chi-Square test to determine whether the data support or refute this theoretical connection between the variables. We'll initially assume that ownership and residency are *not* related, and see if the data provide compelling evidence that they are.

Suppose that one-tenth of the students are commuters. If the null hypothesis is right (commuting and car ownership are unrelated), then one-tenth of the car owners commute, and one-tenth of the non-owners also commute. We understand that the sample results might not show precisely one-tenth for each group, and therefore anticipate some small departures from the theoretical expected values. If, however, the deviations are sufficiently large we will reject the null.

For this test, the hypotheses are:

H_0: Car ownership and residency status are *independent*
H_A: H_0 is false

Stat ➤ Tables ➤ Cross Tabulation... Select the two variables, `OwnCar` and `Res` (in that order). Check **Chi-Square analysis,** and select **Above and expected count.**

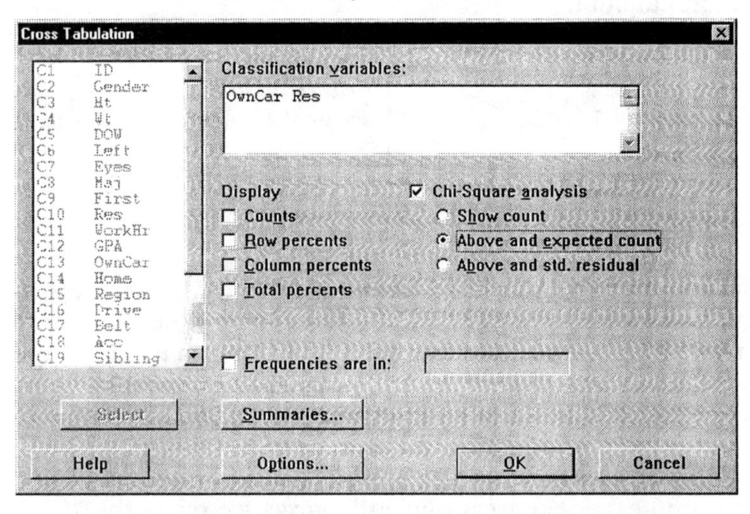

Look at the results in the Session Window. The rows of the table represent car ownership, and the columns represent residency (commuters in first column, residents in the second). The non-owners are in the "N" row, and the car owners are in the "Y" row. Each cell of the table contains an observed joint frequency and an expected frequency.

Thus, for example, there were only 6 commuters who do not own cars; under the null hypothesis, in a sample of 219 students, with 50 commuters and 78 non-owners, 17.81 were expected in that cell.

Tabulated Statistics

```
Rows: OwnCar       Columns: Res
            C          R       All

 N          6         72        78
        17.81      60.19     78.00

 Y         44         97       141
        32.19     108.81    141.00

 All       50        169       219
        50.00     169.00    219.00

Chi-Square = 15.759, DF = 1, P-Value = 0.000
```

The test statistic in this case has a value of 15.759, and under the null hypothesis would follow a Chi-Square distribution with one degree of freedom. Compared to that distribution, the test statistic has a p-value of close to zero, which is small enough to reject the null hypothesis, and conclude that the two variables are *dependent*.

The test for independence comes with one *caveat*—it can be unreliable if the expected count in any cell is less than five. In such cases, a larger sample is advisable. For instance, let's look at another possible relationship in this dataset. One of the questions on the survey asks students to classify themselves as "below average," "average," or "above average" drivers. Let's ask if that variable is independent of gender.

Stat ➤ Tables ➤ Cross Tabulation... Select the two variables, `Drive` and `Gender`. The rest of the dialog remains as is.

Take note of the warning that appears at the end of the Session output shown on the following page. In this instance, the test results may be unreliable due to the uncertainty created by the two cells with very low expected counts. Strictly speaking, we should not come to any inference based upon this sample data. As a description of the sample, though, it is appropriate to note that the men seemed to have higher opinions of their driving than the women do of theirs.

Tabulated Statistics

```
Rows: Drive      Columns: Gender

            F          M        All

A          61         45        106
        46.19      59.81     106.00

AA         30         74        104
        45.32      58.68     104.00

BA          4          4          8
         3.49       4.51       8.00

All        95        123        218
        95.00     123.00     218.00

Chi-Square = 17.727, DF = 2, P-Value = 0.000
2 cells with expected counts less than 5.0
```

Testing for Independence (Summary Data Only)

In the example just given, we started with case-wise data on a sample. Sometimes, we find a cross tabulation in a news report or other literature, and might want to conduct the appropriate test. This is easily done.

For example, suppose we were interested in the marital status of adults in Massachusetts, and found a newspaper report with the following crosstabulation of Gender and Marital status. (The cells here contain frequencies extracted from our Census data file.)

	Male	Female
Never	99	81
Married	215	213
Widowed	15	56
Divorced	33	47

File ➤ New ➤ Minitab Worksheet

Enter the table from the previous page into C1–C3, as shown:

↓	C1-T Status	C2 Male	C3 Female
1	Never	99	81
2	Married	215	213
3	Widowed	15	56
4	Divorced	33	47
5			

Stat ➢ Tables ➢ Chi-Square Test... Specify that the table is in columns C2–C3. The output is very similar to that shown earlier, except for some additional rows at the bottom.

```
Chi-Sq =   2.014 +   1.837 +
           0.579 +   0.528 +
          10.507 +   9.581 +
           0.697 +   0.635 = 26.378
DF = 3, P-Value = 0.000
```

Test statistic

What do we conclude about marital status and gender? How can you explain this result?

Moving On...

Use the techniques of this lab session to respond to these questions. *For each question, explain how you come to your statistical conclusion, and suggest a real-world reason for the result.*

Census90

1. Is the ability to speak English independent of gender?
2. Is job-seeking independent of gender?

Student

3. Is seatbelt usage independent of car ownership?
4. Is seatbelt usage independent of gender?
5. Is travel outside of United States independent of gender?

6. Is belt usage independent of familiarity with someone who has been struck by lightning? ("Zap"—perhaps such first-hand knowledge makes people more conscientious about safety!)

Mendel

This file contains summarized results for another one of Mendel's experiments. In this case, he was interested in four possible combinations of texture (smooth/wrinkled) and color (yellow/green). His theory would have predicted proportions of 9:3:3:1 (i.e., smooth yellow most common, one-third of that number smooth green and wrinkled yellow, and only one-ninth wrinkled green).[1] The first column of the dataset contains the four categories, and the second and fourth columns contain the respective observed and expected frequencies.

7. Perform a Goodness of Fit test to determine whether these data refute Mendel's null hypothesis of a 9:3:3:1 ratio.

8. Renowned statistician Ronald A. Fisher re-analyzed all of Mendel's data years later, and concluded that Mendel's gardening assistant may have altered the results to bring them into line with the theory, since *each one* of Mendel's many experiments yielded Chi-Square tests similar to this and the one shown earlier in the session. Why would so many consistent results raise Fisher's suspicions?

Salem

This file contains the data about the residents of Salem Village during the witchcraft trials of 1692.

9. Are the variables `ProParris` and `Accuser` independent?

10. Are the variables `ProParris` and `Defend` independent?

---

[1] Kohler, Heinz (1994) *Statistics for Business and Economics*, Third edition. (New York: HarperCollins) pp. 458–459.

Session 14

Analysis of Variance

Objectives

In this session, you will learn to do the following:

- Perform and interpret a one-factor Analysis of Variance
- Understand the assumptions necessary for ANOVA procedures to yield reliable results
- Investigate the connection between two-sample t-tests and ANOVA
- Perform and interpret a two-factor ANOVA

Comparing the Means of More than Two Samples

In Session 12, we learned how to perform tests that compare the mean of one population to the mean of another. At the heart of such tests is an approach that focuses on the *difference between two means.* The null hypothesis specifies a value for the difference. In taking this approach, we do not need to specify two distinct mean values, but rather just one value.

Suppose we wanted to compare the means of *three or more* groups; we could not rely on the trick of hypothesizing a difference of zero, because we cannot simply subtract three means from one another. Instead, we take an entirely new approach to the subject, thinking instead about *comparing variation within and among* different sample groups.

A Simple Example

Consider a population of college students, some of whom work part-time to support themselves while in school. We might hypothesize that those who work many hours outside of school would find their grades suffering. To test that theory, we might use some sample data like that found in our **Student** file. Open that worksheet.

One of the variables is called `WorkHr`, and represents the number of hours per week that a student works part-time. Let's create three groups of students: those who don't have a job, those who work fewer than 20 hours, and those who work 20 hours or more.

> **Manip ➤ Code ➤ Numeric to Text...** Code the data from column `WorkHr` into a new column, `WorkCat`, assigning a value of `None` to students who work 0 hours, a value of `Some` to those working 1:19 hours (from 1 through 19), and a value of `Many` to those working 20:99.

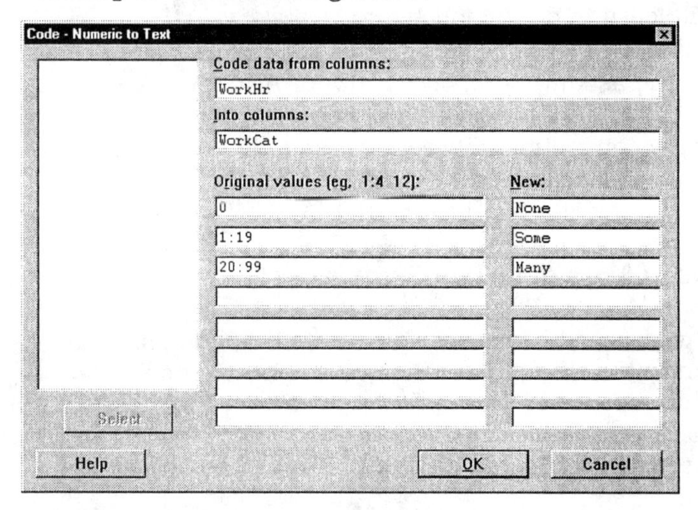

For a first look at a comparison of the average GPA for each of these three new groups, do the following:

> **Stat ➤ Basic Statistics ➤ Display Descriptive Statistics...** Select the variable `GPA`, and **by variable** `WorkCat.` Also, click on **Graphs**, and request a **Boxplot**.

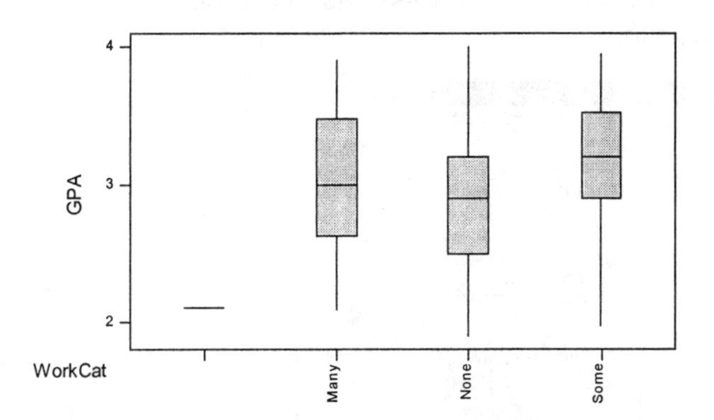

The boxplot shows some variation across the groups, with the highest GPA's belonging to students who work between 1 and 19 hours.[1] As you look in the Session Window, you should notice that the mean GPA differs slightly among the groups, but then again, the sample means of *any* three groups of students would differ to some degree. That's what we mean by sampling error.

Thus, the inferential question is Should we ascribe the observed differences to sampling error, or do they reflect genuine differences among the three populations? Neither the boxplot nor the sample means offer decisive evidence.

Part of the reason we perform a formal statistical test like ANOVA is to clarify some of the ambiguity. The ANOVA procedure will distinguish the extent of the variation we can ascribe to sampling error, and the variation we can ascribe to the *factor*, that is, hours of work.

In this instance, we would initially hypothesize no difference among the three groups. (i.e., The mean GPA for each of the three is identical.) Formally, our null and alternative hypotheses would be:

H_o: $\mu_1 = \mu_2 = \mu_3$
H_A: H_o is false

---

[1] Your descriptive statistics output will show a fourth "group" as well, marked with an asterisk (*). This represents one student who did not indicate how many hours a week s/he works, but whose GPA was reported.

To perform the ANOVA in this case, do the following:

Stat ➤ ANOVA ➤ One-way... The **Response** variable is `GPA`; the **Factor** is `WorkCat`.

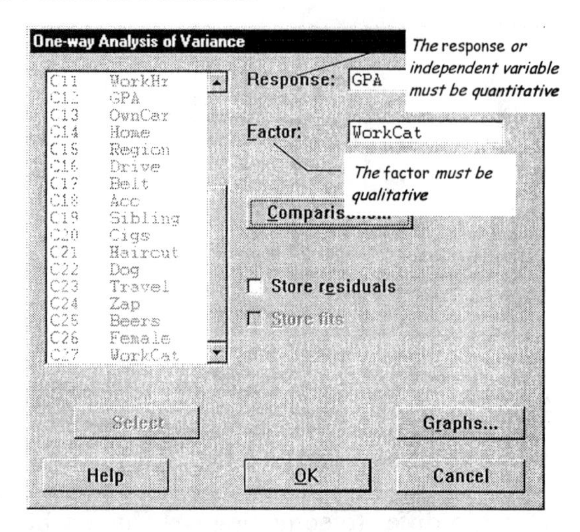

The ANOVA output is best understood as consisting of two parts. The upper portion is the typical ANOVA table, similar to that shown in your textbook. In this instance, you can see that the test statistic (F) equals 8.86, with a corresponding p-value of 0.000. In this test, we would reject the null hypothesis, and conclude that these data do provide substantial evidence of at least one meaningful difference in mean GPA among the three groups of students.

One-way Analysis of Variance

```
Analysis of Variance for GPA
Source     DF        SS        MS        F        P
WorkCat     2     3.558     1.779     8.86    0.000
Error     209    41.943     0.201
Total     211    45.501
                                    Individual 95% CIs For Mean
                                    Based on Pooled StDev
Level       N      Mean     StDev   ----+---------+---------+---------+--
Many       36    3.0197    0.4626         (---------*---------)
None       99    2.8800    0.4584   (-----*-----)
Some       77    3.1664    0.4270                     (------*------)
                                    ----+---------+---------+---------+--
Pooled StDev =   0.4480                2.85      3.00      3.15      3.30
```

The upper part of the output looks like the ANOVA table described in your text.

Subgroup summaries

The lower portion of the ANOVA output shows the sample size, mean, and standard deviations for each of the three subgroups, as well as 95% confidence interval estimates for the three means. Note the extent to which they overlap, reflecting the degree of sampling error here.

Note the "Pooled St. Dev." in the lower left. ANOVA requires three conditions for reliable results. First, the sub-populations should all be *normally* distributed. Second, the *variances* of the populations must be equal, and third, the samples must be *independent* of one another. The pooled standard deviation estimates the square root of the common variance of the three populations.[2]

ANOVA and the Two-Sample t-Tests

Since ANOVA can be used to compare more than two means, you might wonder if it can also compare two means. In fact, it can. A question at the end of Session 12 asked whether car owners had more or fewer accidents than non-owners. At that time, we would have done a 2-sample t-test. The results of that test are $t = 2.10$ and $p \leq 0.037$; our conclusion is that car owners have more accidents than non-owners.

Here is an equivalent ANOVA:

Stat ➢ ANOVA ➢ One-way... Select the variable `Acc` as the response, and `Owncar` as the factor.

Look at the results in the Session Window. ***At a 5% significance level, would you conclude that the two sub-populations share the same mean number of accidents? What is the p-value?***

The p-value is identical to the two-tailed t-test. In a sense, this is not surprising, since these are equivalent tests, being performed with the same data. As such, they should give the same results. What is more, look closely at the two test statistics: $F = 4.39$ and $t = 2.10$. Though the rounded values disguise the fact, t is the square root of F.

Another Example

Above, we alluded to the assumption of equal variances. In the examples so far, we haven't dealt with that assumption. Before relying on

---

[2] Though we didn't do it in this example, we could easily check the normality assumption by looking at the Graphical Summary in the Describe procedure, using the response variable, described *by* the factor variable.

the results of an ANOVA, we should assure ourselves that the three assumptions are reasonable. We already know from earlier labs how to interpret a normality test. The independence of samples is a matter of reasoning and examination of sampling methods. In this example, we'll see one simple way to check for equal variances.

The example concerns the *inventory turnover* among beverage firms. Inventory turnover is a measure of the rate at which a firm's product is sold to consumers. We might wonder if different industry segments experience different inventory turnover. Specifically, we'll ask whether inventory turnover varies by SIC group. We have a numerical response variable (turnover), and a categorical factor (SIC). Open the **Bev** worksheet.

Stat ➤ ANOVA ➤ One-way... The response variable is `Invturn`, and the factor is `SIC`. Before clicking **OK**, click the **Graphs** button. In the dialog, check the box marked **Residuals versus fits.** Click **OK**.

> "**Residuals**" are the differences between the mean value of each SIC group and the actual turnover value for the particular firms.
>
> "**Fits**" are the estimated mean values for each SIC group.

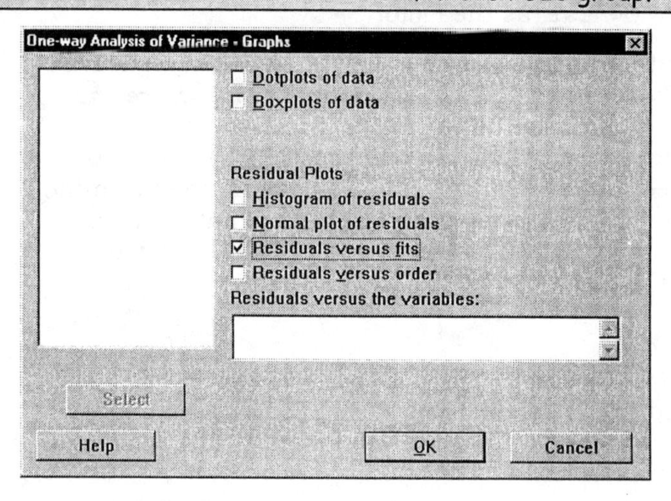

This command graphs the residuals versus mean turnover by SIC code. Note the differences in variation across the groups. If the variances for each group were equal, the residual graph should show vertical clusters of points, with each cluster having about the same range. That is not so here. The two clusters to the left have similar profiles, as do the third and fifth clusters, but those are different from the first two. This suggests that the variances are unequal, and casts doubt on the reliability of the test.

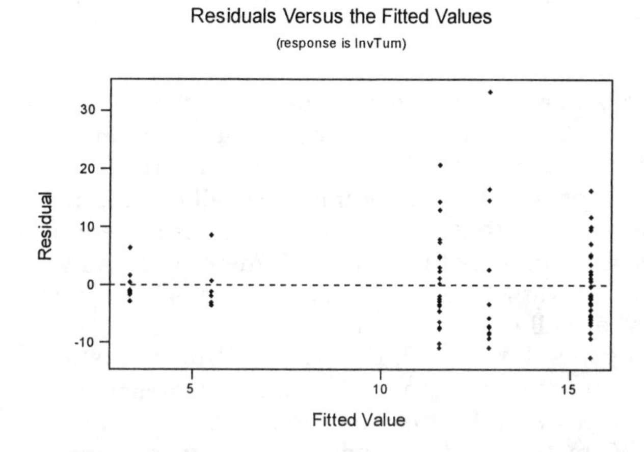

Are the sub-populations normally distributed? As the figure below shows, the entire sample does not look normal, and the normality tests for the five different SIC groups shed doubt on the normality assumption.

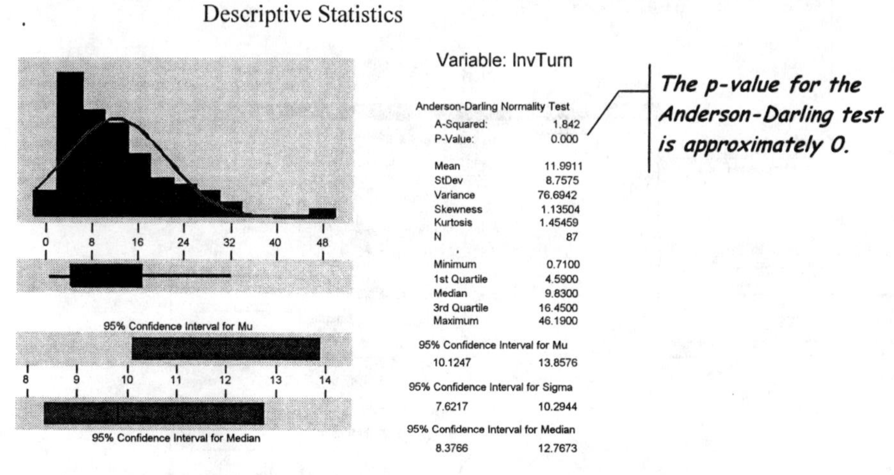

So, in this ANOVA, we seem to have violations of two of our assumptions. As such, even though the F-test indicates that we reject the null hypothesis, in practice we should view this test as inconclusive. There are steps we can take to analyze the data more reliably, but they are beyond the scope of this session. For now, we must conclude that we cannot conclude anything from this particular sample. In Session 20, we'll review some techniques for analyzing data such as these.

Unstacked Data

In the prior examples, all of the data were *stacked* in the worksheet. That is, the response variable values for all groups are in one column, and the factor is in another. Sometimes, we may have the observations for different groups in separate columns. Recall our high school swimmers data. Suppose we want to know if race times are equal among swimmers who specialize in one of the three 50-meter events in our dataset. In other words, we want to know if "back-strokers," "breast-strokers," and "free-stylists" are all equally fast in a 50-meter race.

Open the worksheet called **Swimmer2.** Each row of this worksheet represents a swimmer, and each column (from C5 onward) represents finish times for a different event. We'll first isolate those swimmers who compete in just one of the 50-meter events, by using the copy columns command.

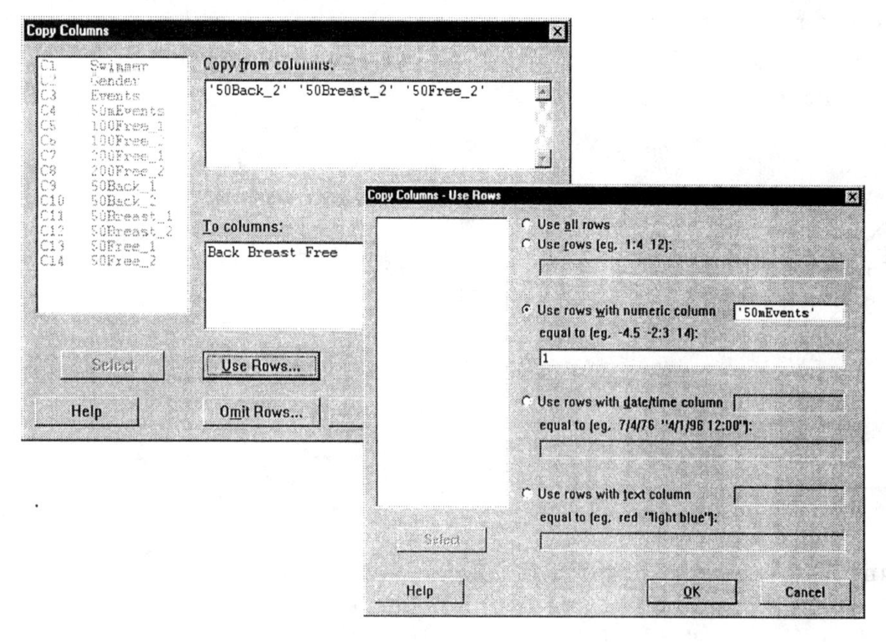

🖰 **Manip ➤ Copy Columns...** The data we want are in columns 10, 12, and 14: `50Back_2`, `50Breast_2`, and `50Free_2`. We will copy from those columns into three new variables, called **`Back`**, **`Breast`**, and **`Free`**. Click **Use Rows,** and specify that we'll use rows where the numeric variable **`50mEvents`** = 1. This will just copy the times for swimmers who competed in one 50-meter event.

In the three new columns, we'll have race times from the second heat in each of the three events. Note that each new column contains a different number of observations, and that *each row no longer represents a single swimmer.*

🖰 **Stat ➤ ANOVA ➤ One-way (Unstacked)...** Select the three new columns.

What do you conclude from the ANOVA results? Can you think of reasons to explain the apparent differences among these three groups of swimmers?

Suppose you learned that half of the free-stylists are female, but the proportions for the other two events are different. Could gender be a factor in race times? Does it matter that not all of the categories are evenly split between genders? Questions like this are better answered with a two-way Analysis of Variance.

A Two-Way ANOVA

In a two-way ANOVA, we hypothesize that there are two categorical factors which might influence the value of the response variable. Thus, we have factor A, which might have *a* distinct levels (values), and factor B, with *b* levels. We can hypothesize that the response variable has equal means for all levels of A, and equal means for all levels of B. In addition, we can hypothesize that the response variable has different means for the *interaction* of A and B. Unlike the single-factor ANOVA, in a two-way analysis, *we must have equal numbers of observations for each of the a x b treatments* or cells. This is an important difference between one- and two-way ANOVA.

Our example comes from the Student data, and the response variable is `Haircut`—the last price the student paid for a professional haircut. In an earlier session, we found that women tend to pay significantly more than men, so we know that Gender is a factor to consider in an analysis of haircut prices. We might also want to consider where these students live. The survey was done on the first day of class, so that a most recent haircut was probably done where students spent the summer. Did students from urban areas pay more than those in rural areas?

In this example, then, we have our numeric response variable (`Haircut`) and two factors (`Sex` and `Region`). For each combination of gender and region, we must observe the same number of students. Since the original file did not have equal numbers of students in each cell, we'll use a different worksheet, with randomly selected data from the original file:

File ➤ Open Worksheet... Choose **Haircut**.

Stat ➤ ANOVA ➤ Two-way... Specify the variables as shown in the completed dialog below, and check that you want to **Display means** for each cell.

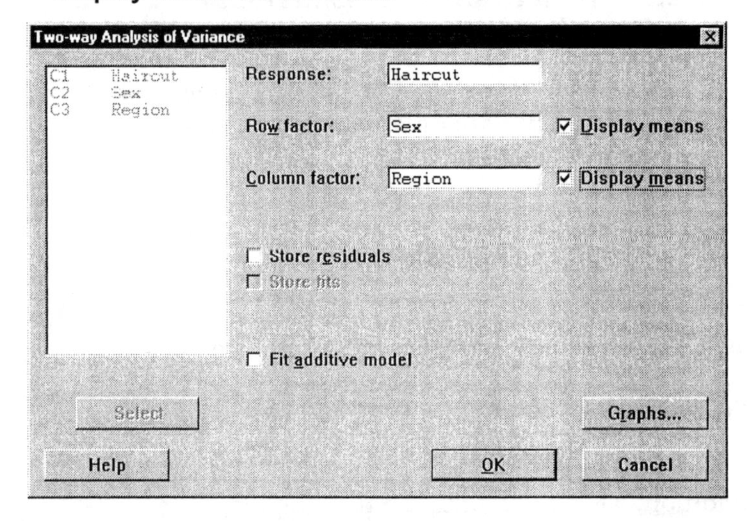

The results in the Session Window are very similar to the one-way ANOVA. In the upper portion, instead of one F ratio, there are now three, corresponding to the two factors and to their interaction. In this example, the p-values for each factor suggest a statistically significant effect on the response variable. As for their interaction, the results are less compelling. The p-value of 0.062 would lead us to reject the null at a 10% significance level, but fail to reject at the 5% level.

What does this mean? It means that women pay different mean prices than men, that urban, suburban, and rural students don't pay the same prices, and that the *extent* of the differences due to gender may vary by region.

Two-way Analysis of Variance

```
Analysis of Variance for Haircut
Source          DF        SS        MS        F        P
Sex              1    2674.0    2674.0    67.38    0.000
Region           2     703.6     351.8     8.86    0.000
Interaction      2     231.8     115.9     2.92    0.062
Error           54    2143.0      39.7
Total           59    5752.4

                       Individual 95% CI
Sex             Mean   -----+---------+---------+---------+------
F               23.5                                (----*----)
M               10.1    (---*----)
                       -----+---------+---------+---------+------
                          10.0      15.0      20.0      25.0

                       Individual 95% CI
Region          Mean   ----------+---------+---------+---------+-
R               13.6    (-------*-------)
S               21.5                            (--------*-------)
U               15.2        (-------*-------)
                       ----------+---------+---------+---------+-
                               14.0      17.5      21.0      24.5
```

Moving On...

Use the techniques of this session to respond to the following questions. **Check the underlying ANOVA assumptions where possible using appropriate graphs; also, explain both the statistical evidence and theoretical reasons for your responses to the questions.**

Bev

1. Do total company assets vary by SIC code?
2. Do gross sales vary by SIC code?
3. Does the amount of revenue per employee vary by SIC code?

Aids

4. Did the 1992 AIDS rates vary significantly by WHO Region? Did the 1993 rates do so?

Nielsen

5. Does the mean rating vary by television network?

Mft

This dataset contains Major Field Test (MFT) results, SAT scores and GPA's for a group of college seniors majoring in a science. Department faculty are interested in whether they can predict a senior's MFT performance based on their high school or college performance. The variables GPAQ, VerbQ, and MathQ indicate the quartile in which a student's GPA, Verbal SAT, and Math SAT scores fall within the sample.

6. Do mean total scores on the MFT vary by GPA quartile? Comment on distinctive features of this ANOVA.
7. Does the relationship between total score and GPA hold true for each individual portion of the MFT?
8. Do mean total scores on the MFT vary by Verbal SAT quartile? Math SAT quartile?
9. Suggest some ways that the department could characterize MFT performance using the data available.
10. Based on these results, one faculty member suggests that College grading policies need revision. Why might one think that, and what do you think of the suggestion?

Falcon

This worksheet contains the results of a study to investigate the residual effects of DDT among falcons. DDT is a pesticide which was banned in the United States due to its long-lasting detrimental effects on bird populations. Years later, residues of DDT were still present in birds.

11. Using a two-way ANOVA, does the evidence suggest a relationship between the amount of DDT residue and the age of the bird? The nesting site? Is there a significant interaction between the two factors?

Session 15

Linear Regression (I)

Objectives

In this session, you will learn to do the following:

- Perform a simple, two variable linear regression analysis
- Test hypotheses concerning the relationship between two quantitative variables
- Evaluate the Goodness of Fit of a linear regression model

Linear Relationships

Some of the most interesting questions of statistical analysis revolve around the relationships between two variables. How much will sales increase if we spend more on advertising? How much will regional water consumption increase if the population increases by 1,000 people? How much more heating fuel will I use if I add a room to my house?

In each of these examples, there are two common elements—a pair of quantitative variables (e.g., sales revenue and advertising expense), and a *theoretical reason* to expect that the two variables are somehow related.

Linear regression analysis is a tool with several important applications. First, it is a way of *testing hypotheses* concerning the relationship between two numerical variables. Second, it is a way of *estimating* the specific nature of such a relationship. Beyond asking, "Are sales and advertising related?" regression allows us to ask *how* they are related. Third, and by extension, it allows us to *predict* values of one variable if we know or can estimate the other variable.

As a first illustration, consider the classic economic relationship between consumption and income. Each additional dollar of income enables a person to spend (consume) more. As such, when income increases, we expect consumption to rise as well. Similarly, a person with more income than I have will, other things being equal, tend to consume more than I do. Let's begin by looking at aggregate income and consumption of all individuals in the United States, over a long period of time.

File ➤ Open Worksheet... Select the file **Us**. This file contains different economic and demographic variables for the years 1965–1996. We are interested in `PersCon` and `PersInc`, which represent aggregate personal consumption and aggregate personal income, respectively.

First, let's construct a scatterplot of the two variables. Our theory says that consumption depends on income. In the language of regression analysis, consumption is the **dependent** or **response** variable. Income is the **independent** or **predictor** variable. It is customary to plot the dependent variable on the Y-axis, and the independent on the X-axis.

Graph ➤ Plot The Y variable is `PersCon` and the X variable is `PersInc`. Click **OK**.

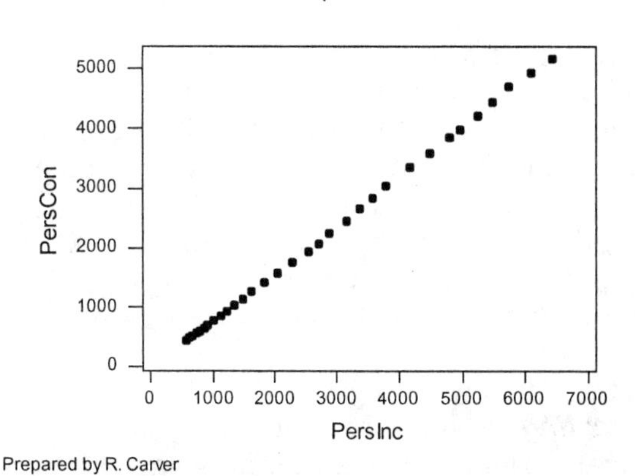

As you look at the resulting plot, you can see that the points fall into nearly a perfect straight line. This is an example of pronounced *positive* or *direct* relationship, and a good illustration of what a linear relationship looks like. It is called a positive relationship because the line has a positive, or upward, slope. One interpretation of the phrase "linear relationship" is simply that X and Y form a line when graphed. But what does that mean in real-world terms? It means that Y changes by a **constant amount** every time X increases by one unit.

In this graph, the pattern formed by the points is nearly a perfect line. The regression procedure will estimate the equation of that line which comes closest to describing the pattern formed by the points.

Stat ➤ Regression ➤ Regression... The **Response** variable is `PersCon`, and the **Predictor** is `PersInc`.

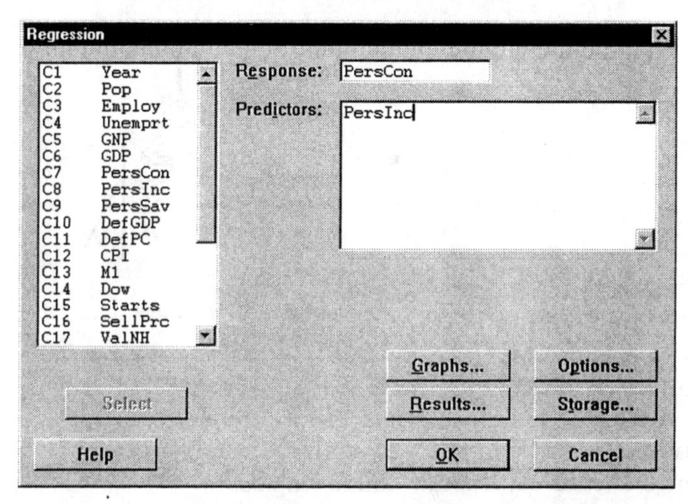

Since the output from the regression procedure is fairly involved, the output from this example is shown on the next page. On your screen, maximize the Session Window, scroll back to the words "Regression Analysis," and follow this discussion while checking your screen.

The regression output consists of five parts: the estimated equation, the table of coefficients, the Goodness of Fit measures, the ANOVA table, and a table of unusual observations. Your text may deal with some or all of these parts in detail; in this session, we'll consider them one at a time.

The uppermost part is the estimated equation. Minitab estimates that the line that most nearly fits these points is given by the equation

```
PersCon = - 38.6 + 0.809 PersInc
```

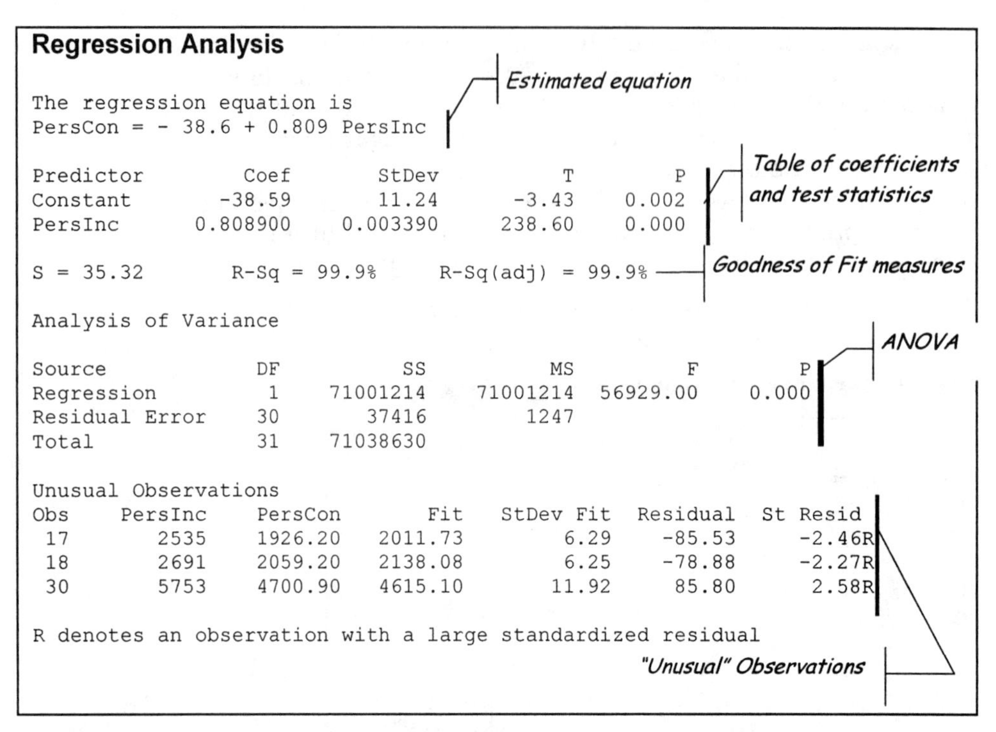

Regression Analysis

```
The regression equation is
PersCon = - 38.6 + 0.809 PersInc

Predictor         Coef        StDev          T         P
Constant        -38.59        11.24      -3.43     0.002
PersInc       0.808900     0.003390     238.60     0.000

S = 35.32        R-Sq = 99.9%      R-Sq(adj) = 99.9%

Analysis of Variance

Source            DF          SS           MS          F        P
Regression         1    71001214     71001214   56929.00    0.000
Residual Error    30       37416         1247
Total             31    71038630

Unusual Observations
Obs     PersInc     PersCon        Fit    StDev Fit   Residual   St Resid
 17        2535     1926.20    2011.73         6.29     -85.53     -2.46R
 18        2691     2059.20    2138.08         6.25     -78.88     -2.27R
 30        5753     4700.90    4615.10        11.92      85.80      2.58R

R denotes an observation with a large standardized residual
```


The slope of the line (.809) means that if Personal Income increases by one billion dollars, Personal Consumption increases by 0.809 billion dollars. In other words, in the aggregate, Americans consumed about 81 cents of each additional dollar earned.[1]

What does the intercept mean? The value of -38.6 literally says that if income were 0, consumption would be -38.6 billion dollars. This makes little sense, but reflects the fact our dataset lies very far from the Y axis. In essence, the estimated Y-intercept is a huge extrapolation beyond the observed data. The line which we estimate must cross the axis somewhere; in this instance, it crosses at -38.6.

Now look at the table of coefficients. Notice that next to the word **`Constant`** is –38.59, and next to **`PersInc`** is 0.808900. These are the same values as in the equation above, except that they are carried to

---

[1] You may recognize this as the *marginal propensity to consume.* Note that the slope refers to the *marginal* change in Y, given a one unit change in X. It is *not* true that we spent 81 cents of *every* dollar. We spend 81 cents of the next dollar we have.

several more decimal places. Because this relationship is so unusually linear, another example will serve better to explain the rest of the output.

Another Example

Besides consuming our income, we can also save or invest it. Let's examine the relationship between Income and Savings, and compare it to the regression we've just done.

Graph ➤ Plot This time, `PersSav` is Y, and `PersInc` is X.

How does this graph compare to the first scatterplot? Does there appear to be any kind of relationship? In general, are low X values associated with low or high Y values? Are high X values associated with low or high Y values?

Clearly, the connection between Savings and Income is not nearly as strong as the connection between Consumption and Income. To measure the strength of relationships, we return to the correlation coefficient.

Stat ➤ Basic Statistics ➤ Correlation... Select the variables `PersSav` and `PersInc`.

The correlation between Consumption and Income is approximately 1.0. ***What does it mean to say that Savings and Income have a correlation of less than 1?***

Stat ➤ Regression ➤ Regression... Now, the **Response** is `PersSav`, and the **predictor** is the same as before. ***Interpret the resulting equation. What does the slope of the line tell you about savings and income?***

Inference from Output

If we were to hypothesize a relationship between Income and Savings, it would be positive: the more you earn, the more you can save. Formally, the theoretical model of the relationship might look like:

Savings = Intercept + (slope)(Income) + ε or

$$Y = \beta_0 + \beta_1 X + \varepsilon$$

If X and Y genuinely have a positive relationship, β_1 is a positive number. If they have a negative relationship, β_1 is a negative number. If they have *no relationship at all,* β_1 is zero.

When we estimate the line using the regression procedure, we compute an estimated slope. Typically, this slope is non-zero. It is critical to recognize that the estimated slope is a result of the particular sample at hand. A different sample would yield a different slope. Thus, our estimated slope is subject to sampling error, and therefore, is a matter for hypothesis testing.

One of the standard tests we perform in a regression analysis is designed to judge whether there is any *significant linear relationship* between X and Y. Our null hypothesis is that there is none. (i.e., that the 'true' slope is zero)

$$H_0: \beta_1 = 0$$
$$H_A: \beta_1 \neq 0$$

```
The regression equation is
PersSav = 41.0 + 0.0364 PersInc

Predictor          Coef         StDev            T        P
Constant         40.980         9.697         4.23    0.000
PersInc        0.036379      0.002924        12.44    0.000
```

Now look at the table of coefficients for this regression. The right-most two columns are labeled T and p. These represent t-tests asking if the intercept and slope (respectively) are equal to zero.[2] In this case, the value of the test statistic for the slope is 12.44, and the p-value associated with that test statistic is approximately 0. As in all t-tests, we take this to mean that we should *reject* our null hypothesis, meaning *there is a statistically significant relationship* between X and Y.

We know that the regression procedure, via the *least squares method* of estimation, gives us the line which fits the points better than any other. We might ask just how "good" that fit is. It may well be the case that the "best fitting" line is not especially close to the points at all!

Another standard part of regression output is a statistic which addresses just that question. The statistic is called the ***coefficient of determination,*** represented by the symbol r^2. It is the square of r, the coefficient of correlation. Among the Goodness of Fit measures in the

[2] In this example, the intercept has little practical meaning. Therefore, in this session, we bypass it. A later example discusses a hypothesis test concerning the intercept.

regression output, locate **R-sq.**[3] r^2 can range from 0 to 100%, and indicates the extent to which the line fits the points; 100% is a perfect fit, such that each point is on the line. The higher the value of r^2, the better.

```
S = 30.46          R-Sq = 83.8%          R-Sq(adj) = 83.2%
```

To better visualize how this line fits the points, do the following:

Stat ➤ Regression ➤ Fitted Line Plot... Select **PersSav** as the **Response**, **PersInc** as the **Predictor**. Maximize the resulting graph to get a good look at the graph. According to these results, about 84% of the variation in Personal Saving is associated with changes in Personal Income.

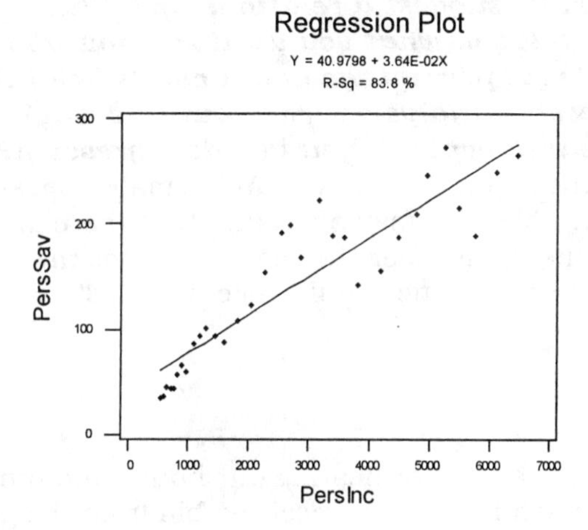

An Example of a Questionable Relationship

We've just seen two illustrations of fairly strong, statistically significant linear relationships. Let's look at another example, where the theory is not as compelling.

Nutrition and session experts concern themselves with a person's body fat, or more specifically the percentage of total body weight made up of fat. Measuring body fat precisely is more complicated than measuring other body characteristics, so it would be nice to have a mathematical

---

[3] You will also find R-sq (adj) next to R-sq. The *adjusted* r^2 is used in Multiple Regression, and we will discuss it in a later lab session.

model relating body fat to some easily measured human attribute. Our dataset called **Bodyfat** contains precise body fat and other measurements of a random sample of adult men. Open the worksheet.

Suppose we wondered if height could be used to predict the percentage of body fat. We could use these data to investigate that relationship.

> **Graph ➤ Plot** Y is `FatPerc` and X is `Height`. ***Is there evidence in this graph of a positive linear relationship?***

> **Stat ➤ Regression ➤ Regression...** `FatPerc` is the response, and `Height` is the predictor.

Do the regression results suggest a relationship? What specific parts of the output tell you what you want to know about the connection between body fat percentage and a man's height? Why do you think the regression analysis turned out this way? What, if anything, does the intercept tell you in this regression?

The key point here is that even though we can estimate a least squares line for any pair of variables, we may often find that there is no statistical evidence of a relationship. Neither the scatterplot nor the estimated slope are sufficient to determine significance; we must consult a t- or F-ratio for a definitive conclusion.

An Estimation Application

In my home, we use natural gas for heating our house and our water, as well as for cooking. Each month, we receive a bill from the gas company for the gas we burn. The bill is a treasure trove of information, including two variables which are the focus of this example.

The first is a figure which is approximately equal to the number of cubic feet of natural gas consumed per day during the billing period. More precisely, it equals the number of "therms" per day; a therm is a measure of gas consumption reflecting the fact that the heating capacity of natural gas fluctuates during the year.

The second variable is simply the mean temperature for the period. These two variables are contained in the data file called **Utility**.

> **File ➤ Open Worksheet... Utility** The variables we want are called `GaspDay` and `MeanTemp`.

We start by thinking about *why* gas consumption and outdoor temperature should be related. ***What would a graph of that***

relationship look like? Do we expect it to be linear? Do we expect it to be positive or negative? Before proceeding, sketch the graph you expect to see.

Graph ➤ Plot Construct a scatterplot with `GaspDay` on the vertical axis, and `MeanTemp` on the horizontal. ***Does there appear to be a relationship?***

Stat ➤ Regression ➤ Regression... This time, you decide which variable is the response and which is the predictor.

Now look at the regression results. ***What do the slope and intercept tell you about the estimated relationship? What does the negative slope indicate? Is the estimated relationship statistically significant? How would you rate the Goodness of Fit?***

One fairly obvious use for a model such as this one is to predict or estimate how much gas we'll use in a given month. For instance, in a month averaging temperatures of 40 degrees, the daily usage could be computed as

GaspDay = 14.2 - 0.200 (40)
= 14.2 - 8.0
= 6.2 therms per day.

Use the model to estimate daily gas usage in a month when temperatures average 75°. Does your estimate make sense to you? Why does the model give this result?

A Classic Example

Between 1595 and 1606 at the University of Padua, Galileo Galilei (1564–1642) conducted a series of famous experiments on the behavior of projectiles. Among these experiments were observations of a ball rolling down an inclined ramp (see diagram below). Galileo varied the height at which the ball was released down the ramp, and then measured the horizontal distance which the ball traveled.

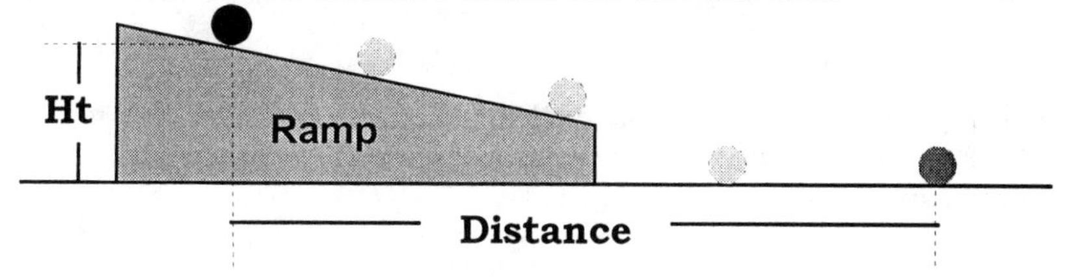

We'll begin by looking at the results of one of Galileo's experiments; the data are in **Galileo**. As you might expect, balls released at greater heights traveled longer distances. Galileo hoped to discover the relationship between release height and horizontal distance. Both the heights and distances are recorded in *punti* (points), a unit of distance.

- First, plot the data in the first two columns of the worksheet, with horizontal distance as the Y variable.

Does the graph suggest that distance and height are related? Is the relationship positive or negative? For what physical reasons might we expect a non-linear relationship?

Although the points in the graph don't quite fall in a straight line, let's perform a linear regression analysis for now.

- Perform the regression, using `DistRamp` as the response.

Using the regression results, comment on the meaning and statistical significance of the slope and intercept, as well as the Goodness of Fit measures. Use the estimated regression equation to determine the release height at which a ball would travel 520 punti.

As we did earlier, let's see the estimated regression line in relation to the sample points:

- **Stat ➤ Regression ➤ Fitted Line Plot...**

Comment on the extent to which the line appears to "fit" the plotted points. Do you think your estimated release height for a 520 punti travel is probably high or low? Explain.

It should be clear that a linear model is not the very best choice for this set of data. Regression analysis is a very powerful technique, which is easily misapplied. In upcoming sessions, we'll see how we can refine our uses of regression analysis to deal with problems such as non-linearity, and to avoid abuses of the technique.

Moving On...

Use the techniques and information in this session to answer the following questions. Explain or justify your conclusions with appropriate graphs or regression results.

Galileo

1. Galileo repeated the rolling ball experiment with slightly different apparatus, described in Appendix A. Use the data in the third and fourth columns of the worksheet to estimate the relationship between horizontal distance and release height.
2. At what release height would a ball travel 520 *punti* in this case?

Us

Investigate possible linear relationships between the following pairs of variables. ***In each case***, comment on (a) why the variables might be related at all, (b) why the relationships might be linear, (c) the interpretation of the estimated slope and intercept, (d) the statistical significance of the model estimates, and (e) the goodness of fit of the model. (In each pair, the Y variable is listed first.)

3. Cars in use vs. Population
4. FedRecpt vs. PersInc.
5. GDP vs. Employ

Mft

These are the Major Field Test scores, with student GPA and SAT results.

Investigate possible linear relationships between the following pairs of variables. ***In each case***, comment on (a) why the variables might be related at all, (b) why the relationships might be linear, (c) the interpretation of the estimated slope and intercept, (d) the statistical significance of the model estimates, and (e) the goodness of fit of the model. **(In each question, the Y variable is the Total MFT score.)**

6. GPA
7. Verbal SAT
8. Math SAT

Bodyfat

9. These are the body fat and other measurements of a sample of men. Our goal is to find a body measurement which can be

used reliably to estimate body fat percentage. For each of the three measurements listed here, perform a regression analysis. Explain specifically what the *slope* of the estimated line means in the context of body fat percentage and the variable in question. Select the variable which you think is best to estimate body fat percentage.

- Chest circumference
- Abdomen circumference
- Weight

10. Consider a man whose chest measurement is 95 cm, abdomen is 85 cm, and who weighs 158 pounds. Use your best regression equation to estimate this man's body fat percentage.

Session 16

Linear Regression (II)

Objectives

In this session, you will learn to do the following:

- Validate the assumptions for least squares regression by analyzing the *residuals* in a regression analysis
- Use an estimated regression line to estimate or predict Y values

Assumptions for Least Squares Regression

In the prior session, we learned how to fit a line to a set of points. Minitab uses a common technique, called the *method of least squares.*[1] Though there are several other alternative methods available, least squares estimation is by far the most commonly used.

We can apply the technique to any set of paired (x, y) values and get an estimated line. However, if we plan to use our estimates for consequential decisions, we want to be sure that the estimates are unbiased and otherwise reliable. The least squares method will yield unbiased, consistent, and efficient[2] estimates when certain conditions are true. Recall that the basic linear regression model states that X and Y

---

[1] This method goes by several common names, but the term "least squares" always appears, referring to the criterion of minimizing the sum of squared deviations between the estimated line and the observed Y values.

[2] You may recall the terms *unbiased, consistent* and *efficient* from earlier in your course. This is a good time to review these definitions.

have a linear relationship, but that any observed (x,y) pair will randomly deviate from the line. Algebraically, we can express this as:

$$y_i = \beta_0 + \beta_1 x_i + \varepsilon_i$$

where

x_i, y_i represent the ith observation of x and y, respectively,
β_0 is the intercept of the underlying linear relationship,
β_1 is the slope of the underlying linear relationship, and
ε_i is the ith random disturbance [i.e., the deviation between the theoretical line and the observed value (x_i, y_i)]

For least squares estimation to yield reliable estimates of β_0 and β_1, the following must be true about ε, the random disturbance.[3]

- *Normality*: At each possible value of *x*, the random disturbances are normally distributed; $\varepsilon | x_i$ follows a normal distribution.
- *Zero mean*: At each possible value of *x*, the mean of $\varepsilon | x_i$ is 0.
- *Homoskedasticity:* At each possible value of *x*, the variance of $\varepsilon | x_i$ equals σ^2, which is constant.
- *Independence*: At each possible value of *x*, the value of $\varepsilon_i | x_i$ is independent of all other $\varepsilon_j | x_j$.

If these conditions are not satisfied and we use the least squares method, we run the risk that our inferences—the tests of significance and any confidence intervals we develop—will be misleading. Therefore, it is important to verify that we can reasonably assume that x and y have a linear relationship, and that the four above conditions hold true. The difficulty lies in the fact that we cannot directly observe the random disturbances, ε_i, since we don't actually know the location of the "true" regression line. In lieu of the disturbances, we instead examine the *residuals*—the differences between our estimated regression line and the observed y values.

Examining Residuals to Check Assumptions

By computing and examining the residuals, we can get some idea of the degree to which the above conditions apply in a given regression analysis. We will adopt slightly different analysis strategies depending on

[3] Some authors express these as assumptions concerning $y | x_i$.

whether the sample data are cross-sectional or time series. In cross-sectional data, the assumption of independence is not especially relevant, since the observations are not made in any meaningful sequence; in time series data, though, the independence assumption is important. We will start with a cross-sectional example.

File ➤ Open Worksheet... States This is the worksheet with data about the 50 states in 1994. Let's first check for a linear relationship between the number of cars insured in a state in 1994 and the population of the state that year.

Graph ➤ Plot `CarsIns` is Y and `Pop` is X.

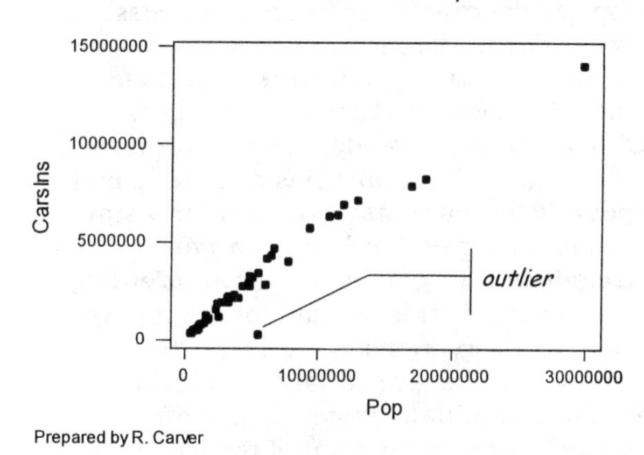

Based on this graph, would you say there is a linear relationship between the two variables? Next we perform the regression, and have Minitab compute and store the estimated Y values and the residuals in the worksheet. There are several ways to accomplish this; here is one way.

Stat ➤ Regression ➤ Regression... Select `CarsIns` as the **Response** variable and `Pop` as the **Predictor**. Before clicking **OK**, click on **Storage....** This button opens another dialog box, allowing you to specify various values for Minitab to compute and store. Check **Fits** and **Residuals**, and then click **OK**. Then click **OK** in the Regression dialog.

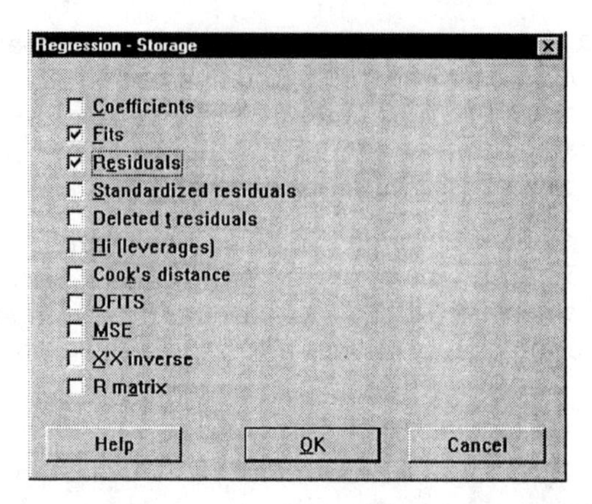

We interpret the regression results exactly as in the prior session. This regression model looks quite good: the significance tests are impressive, and the coefficient of determination (r^2) is quite high. Before examining the residuals *per se*, note the lower section of the output, labeled "Unusual Observations." Three points are identified as unusual. The first has a very large negative residual. This state has a rather small population of 5,500,431 and especially few cars insured, even for a state that size. It lies well below the estimated regression line. ***Can you identify this point in the scatterplot? Can you surmise the identity of the state?*** See the "Moving On" section of this session for the answer.

Each of the other two unusual observations is marked with an "X," indicating an "observation whose X value gives it large influence." Note that these cases do *not* have large residuals. Instead, they are unusual because they exert "leverage" in the estimation of the slope. The two points in question are the right-most points in the graph. The outlying point furthest to the right represents California, and to its left is New York. Because of their extreme X positions, the slope computation depends heavily on these two values; small differences in the Y value of either could have had substantial impact on the estimated slope.

In the Data Window, now scroll to the right. Notice that two new columns have appeared, labeled FITS1 and RESI1, containing the estimated Y values for each state, and the residual for each. We can examine the residuals to help evaluate whether any of the four least squares assumptions should be questioned. We can inspect the residuals as surrogates for the random disturbances.

Recall that there are four assumptions. In cross-sectional data like this, we need not concern ourselves with the fourth (independence), as noted earlier. Of the other three, we are unable to use the residuals to evaluate the *zero mean* assumption, since the least squares method guarantees that the mean of the residuals will be zero.

However, we can examine the residuals to help decide if the other two assumptions—*normality* and *homoskedasticity*—are valid. The simplest tools for doing so are some graphs.

Stat ➤ Regression ➤ Residual Plots... In the dialog, select `RESI1` in the **Residuals** box, and `FITS1` in the **Fits** box. This command generates a group of four graphs (see next page).[4]

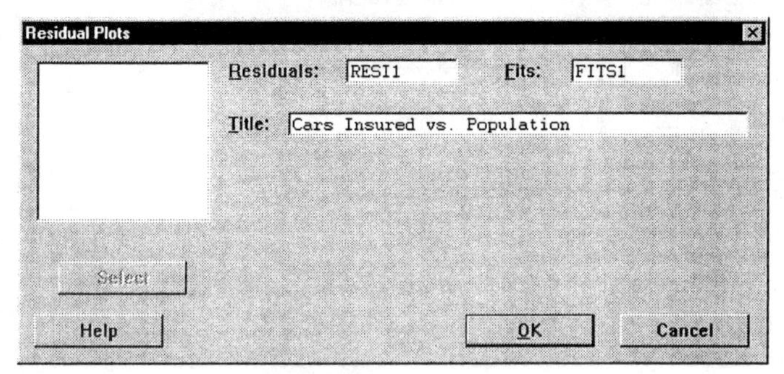

The two graphs on the left side relate to normality. For residuals which are normally distributed, the Normal Probability Plot will look like a 45° upward sloping diagonal line. The Histogram below it should look like a symmetrical, bell-shaped normal distribution. To the extent that the graphs deviate from these patterns, the normality assumption should be questioned.[5] In this case, apart from one outlier (Unusual Observation # 2), the residuals do appear to be normally distributed.

The graph in the upper right, the I-Chart of Residuals, is a sequential plot of the residuals, ordered by observation number. Mostly, this graph is relevant to time series data. Since the order of observations here is arbitrary (alphabetical by state name), we'll skip this graph for now, and discuss it in the next example.

---

[4] Most standard statistics textbooks discuss the interpretation of residual graphs. Consult your text for a more detailed discussion.

[5] For a more rigorous test of normality, run a Graphical Summary of the variable RESI1, using the Descriptive Statistics dialog, and consult the Anderson-Darling Normality test. The null hypothesis is that the data are normally distributed; if the p-value is very small, we would reject that hypothesis.

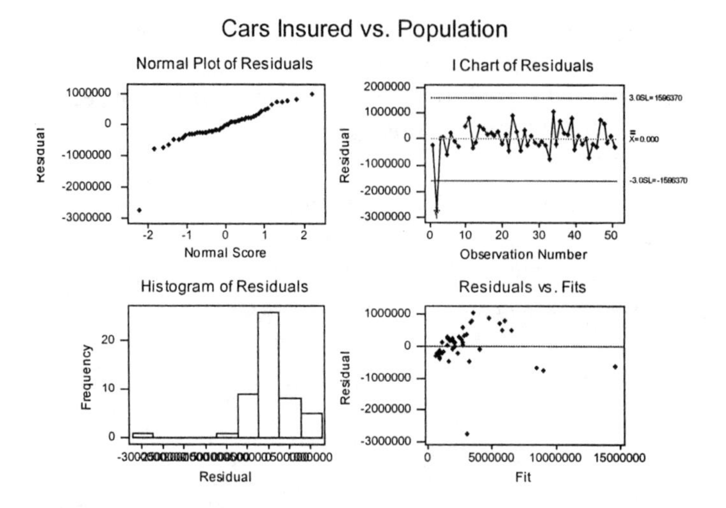

The graph in the lower right is a plot of the residuals versus the fitted, or estimated, values[6]. This graph can give us insight into the assumption of equal variances (homoskedasticity), as well as the assumption that X and Y have a linear relationship. When both are true, the residuals will be randomly scattered in an even, horizontal band around a residual value of zero. Residuals that "fan out" from left to right, or right to left signal *heteroskedasticity*, or unequal variances. A curved pattern suggests a non-linear relationship. Here, we see a generally curved swath of residuals; the ones to the left and right tend to be negative, and those in the center are positive. This suggests that the estimated line passes above the left-most points, below the central points, and above the right-most. In other words, the observed *x-y* values curve around the estimated regression line. If you look back at your scatterplot, you can see a slight bend in the points.

A Time-Series Example

Earlier, we noted that the assumption of independence is often a concern in time series datasets. If the disturbance at time $t + 1$ depends on the disturbance at time t, inference and estimation present some special challenges. Once again, our initial inclination will be to assume that the random disturbances *are* independent, and look for compelling

---

[6] Some authors prefer to plot residuals versus x values; the graphs are equivalent..

evidence that they are not. As before, we do so by running a least squares regression, saving the fitted and residual values, and examining the residuals.

Our next example uses time-series data, so that the sequence of these observations is meaningful. As in the prior session, we'll look at some annual data from the US economy, and return to the relationship between aggregate personal savings and aggregate personal income. Open the file **Us**. As in the last lab, we'll run a regression with `PersSav` as the Y variable and `PersInc` as the X variable.

Stat ➤ Regression ➤ Regression... The response is `PersSav`, and the Predictor is `PersInc`. As in the prior regression, we want to store fits and residuals.

Stat ➤ Regression ➤ Residual Plots... Again, you must indicate the location of the fitted values and residuals in the worksheet. Since we have opened a different worksheet than in the previous example, specify `RESI1` and `FITS1` again. Give this plot an appropriate title.

This time, because the sequence of residuals *is* meaningful, we should examine the "I-chart" of residuals. What are we looking for? If the residuals are independent, then a negative residual at time *t* should not affect the likelihood of a positive or negative residual at time *t+1*. Thus, if we see positive or negative residuals grouped sequentially, we might doubt this assumption. In this chart, do you see areas where a consecutive series of residuals appear above or below the horizontal 0 (zero) line?

Also, note that there are small numbers next to some of the points on the I-Chart. Those refer to specific problems with the residuals, each of which is explained in the Session Window. These specific tests refer not only to the assumption of independence, but raise some questions about normality and homoskedasticity.

Now look back at the residual plots. ***What do you conclude about the assumptions of normality and homoskedasticity?***

Issues in Forecasting and Prediction

One reason that we are concerned about the assumptions is that they affect the reliability of estimates or forecasts. To see how we can use Minitab to make and evaluate such forecasts, we'll turn to another example. Open the file called **Utility**.

This file contains time-series data about the consumption of natural gas and electricity in my home. Also included is the mean monthly temperature for each month in the sample. As in the prior session, we'll model the relationship between gas consumption and temperature, and then go on to forecast gas usage in a month when mean temp is 21 degrees.

As before, we will "plug" 21 degrees into our estimated regression model to get a point estimate. However, we can also develop either a Confidence or a Prediction interval, depending on whether we wish to estimate the *mean* gas use in all months averaging 21 degrees, or the actual gas use in one particular month averaging 21 degrees.

Stat ➤ Regression ➤ Regression As before, **GaspDay** is the response, **MeanTemp** is the predictor. Before clicking OK, we want to select **Options...** and **Storage...**.

Options... In the options dialog, type 21 in the box marked **Prediction intervals for new observations**, to generate confidence and prediction intervals. Click **OK**.

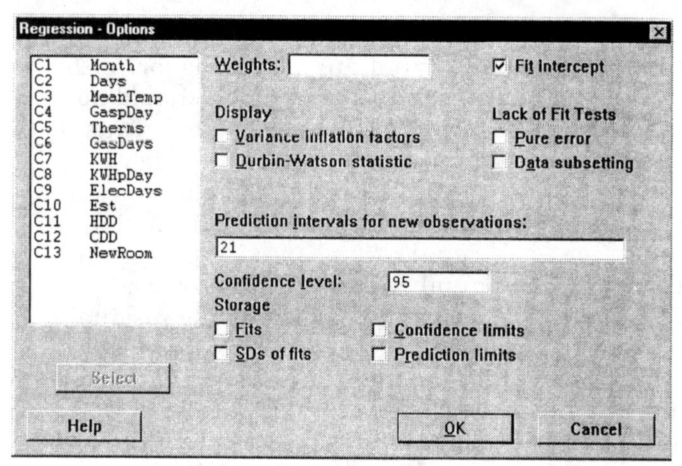

Storage... As before, we want to store both the Fits and Residuals. Click **OK**, and **OK** to the regression command.

```
Predicted Values

     Fit  StDev Fit        95.0% CI              95.0% PI
  10.812      0.305   ( 10.202,  11.421)  (  8.120,  13.503)
```

Now look at the regression results in the Session Window. Below the "unusual observations," you'll see a line displaying the point estimate of `GaspDay` as well as a 95% confidence and prediction interval. ***What do these intervals tell you about natural gas usage when the average daily temperature is 21 degrees?*** Before relying too heavily on this estimate, let's look at the residuals.

Stat ➤ Regression ➤ Residual Analysis... [By now, you should know what to do with this dialog.]

Remember that these are time-series data. ***For each of the three assumptions we can investigate, do you see any problems? How might these problems affect our prediction in this instance?*** To get a better look at the problem here, let's re-plot the data, superimposing our estimated regression line.

Stat ➤ Regression ➤ Fitted Line Plot... In this dialog, enter **`GaspDay`** for the response variable, and **`MeanTemp`** for the predictor.

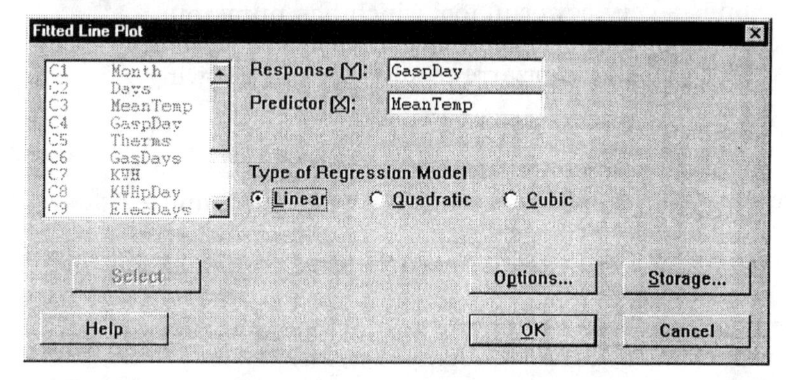

Along the X axis, visually locate 21 degrees, and look up at the regression line and at the observed data points. Note that most of the observed data at such low temperatures lies *above* the regression line, and that the points gently arc around the line. Therefore, the estimated value for 21 degrees is probably too low. What can we do about that? We'll see some solutions in a future lab.

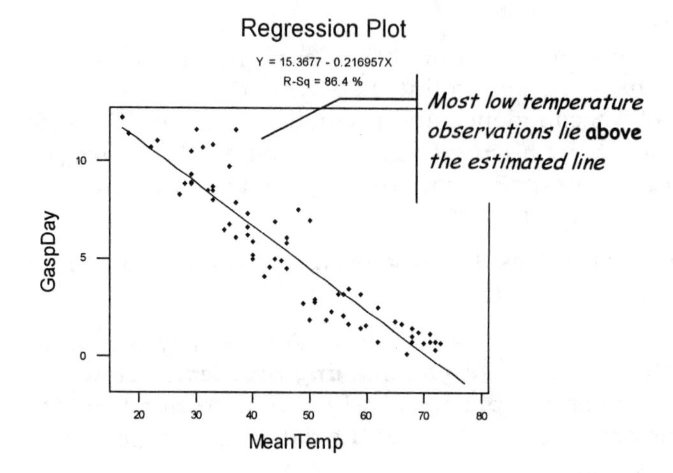

A Caveat about "Mindless" Regression

Linear regression is a very powerful tool which has numerous uses. Like any tool, it must be used with care and thought, though. To see how thoughtless uses can lead to bizarre results, try the following.

File ➤ Open Worksheet... Anscombe

This dataset contains eight columns, representing four sets of X-Y pairs. We want to perform four regressions using X1 and Y1, X2 and Y2, etc. By now, you should be able to run four separate regressions. ***After you do so, look closely at the four sets of results, and comment on what you see.***

Based on the regressions, it is tempting to conclude that the four *x-y* pairs all share the same identical relationship. Or is it?

Graph ➤ Plot Construct four scatterplots (Y1 vs. X1, Y2 vs. X2, etc.) What do you see? Remember that each of these four plots led to the four virtually identical regressions.

Moving On...

Using the techniques of this session, perform and evaluate regressions and **residual analyses** to investigate the following relationships.

Us

Use your regressions and Minitab to predict the dependent variable as specified in the question. In each instance, report the estimated regression equation, and explain the meaning of the slope and intercept.

1. Cars vs. Pop (predict when Pop = 245,000 i.e., 245 million)
2. FedRecpt vs. PersInc (predict when PersInc = 5000 i.e., 5 trillion dollars)

States

Note: The "unusual" state from the first example is Alaska. In the worksheet, enter an asterisk (missing data) in the `Pop` column for Alaska. Now re-do the first example from this session, omitting the unusual case of Alaska.

3. How, specifically, does this affect (a) the regression and (b) the residuals? Compare the slope and intercepts of your two regressions, and comment on what you find.

Bodyfat

4. In Session 15, you did some regressions using these data. This time, perform a linear regression and residual analysis using **`FatPerc`** as the response variable, and **`Weight`** as the predictor. Estimate a fitted value when weight = 158 pounds.
5. Do these sample data suggest that the least squares assumptions have been satisfied in this case?
6. What is the 95 percent interval estimate of mean body fat percentage among all men who weigh 158 pounds?

7. What is the 95 percent interval estimate of body fat percentage for one *particular* man who weighs 158 pound?

Galileo

8. In the previous session, we noted that horizontal distance and release height (first two columns) did not appear to have a linear relationship. Re-run the regression of distance (Y) versus height, and construct the residual plots. Where in these plots do you see evidence of non-linearity?

9. Repeat the same with columns 3 and 4. Is there any problem with linearity here?

Session 17

Multiple Regression

Objectives

In this session, you will learn to do the following:

- Improve a regression model using multiple regression analysis
- Interpret multiple regression coefficients
- Diagnose and deal with multicollinearity

Going Beyond a Single Explanatory Variable

In our previous sessions using simple regression, we examined the relationship between aggregate savings and aggregate income in the United States. In that case, we found a statistically significant relationship between the two variables, but also noted that much of the variation (about 16%) in savings remained *unexplained* by the single variable of income. In addition, the standard error of the estimate (s) was approximately 30 billion dollars.

There are many instances in which we can posit that one variable depends on several others: a single effect with multiple causes. The statistical tool of *multiple regression* enables us to identify those variables simultaneously associated with a dependent variable, and to estimate the separate and distinct influence of each variable on the dependent variable.

For example, in the case of savings, we might hypothesize that aggregate savings in a year depends both on aggregate income and on the prevailing interest rates that year. In our US dataset, we have one variable which measures interest rates: It is called **`NHMort`**, and represents the mean interest rate charged for New Home Mortgages in the observation year. This is different from the interest rates for savings, but does tend to rise and fall with those rates. As such, it is a reasonable *surrogate variable* to use in this example. Let's see if `NHMort` and `Savings` are related. Open **Us**. Since we are interested in relationships among three variables, a **matrix plot** is a good tool to use.

> **Graph ➤ Matrix Plot...** Specify the variables `PersSav`, `PersInc`, and `NHMort`.

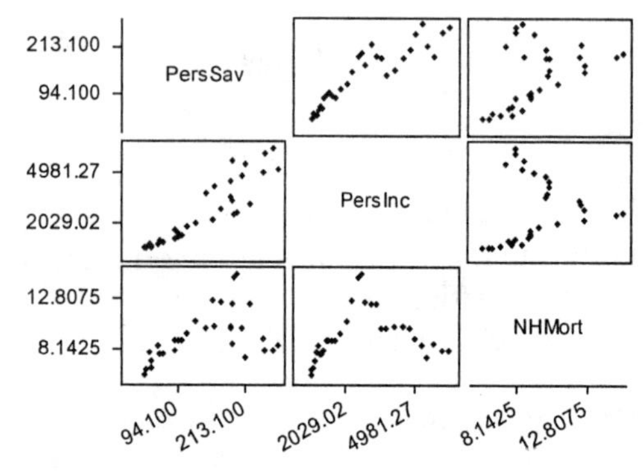

In the resulting plot, we see scatterplots relating each pairing of these three variables. In the first row, both of the graphs have `PersSav` on the Y-axis; in the first column, `PersSav` forms the X-axis. You should recognize the plot of Savings vs. Income. ***What do you see in the plot of savings and interest rates?***

We can also think about these pairings in terms of the correlations between the variables. To compute the correlation coefficients for each pair, do the following:

> **Stat ➤ Basic Statistics ➤ Correlation...** Here again, specify the variables `PersSav`, `PersInc`, and `NHMort`.

In the Session Window, you will find the **correlation matrix,** which reports the correlation between each pairing of the three variables. The matrix is a triangular arrangement of columns and rows; the correlation coefficient for a pair of variables appears at the intersection of the corresponding row and column. For example, savings and income have a correlation of 0.915, while savings and mortgage rates have a correlation of only 0.464. Both correlations are positive, suggesting (for example) that interest rates and aggregate savings rise and fall together, though imperfectly.

Correlations (Pearson)

```
        PersSav  PersInc
PersInc   0.915
          0.000

NHMort    0.464    0.194
          0.007    0.287

Cell Contents: Correlation
               P-Value
```

It is important to recognize that these correlations refer only to the pair of variables in question, without regard to the influences of other variables. Above, though, we theorized that savings varies *simultaneously* with income and interest rates. That is to say, we suspect that interest rates affect savings in the context of a given income level. Therefore, we can't merely look at the relationship of interest rates and savings without taking income into account. Multiple regression allows us to do just that. Let's see how.

Our model in this example is this:

$$Savings_i = \beta_0 + \beta_1 Income_i + \beta_2 NHMort_i + \varepsilon_i$$

We can estimate the coefficients in this model using the regression procedure, much as before.

Stat ➤ Regression ➤ Regression... Select **PersSav** as the **Response** variable, and both **PersInc** and **NHMort** as **Predictors**.

Look in the Session Window (see next page) for the Regression results. They should look quite familiar, with the only differences being an additional line in the table of coefficients, corresponding to **NHMort**, and some new output below the ANOVA table. Look at the coefficients.

We now have one intercept (Constant) and two slopes, one for each of the two explanatory variables. The intercept represents the value of Y when *all* of the X variables are equal to zero. Each slope represents the marginal change in Y associated with a one-unit change in the corresponding X variable, *if the other X variable remains unchanged.* For example, if personal income were to increase by one billion dollars, and mortgage rates were to remain constant, then savings would increase by .0341 billion dollars. Look at the coefficient for `NHMort`. ***What does it suggest?***

```
Regression Analysis

The regression equation is
PersSav = - 43.7 + 0.0341 PersInc + 9.66 NHMort

Predictor          Coef        StDev          T        P
Constant         -43.74        16.42      -2.66    0.012
PersInc        0.034083     0.002088      16.33    0.000
NHMort            9.663        1.704       5.67    0.000

S = 21.33       R-Sq = 92.3%      R-Sq(adj) = 91.8%

Analysis of Variance

Source            DF          SS          MS         F        P
Regression         2      158238       79119    173.84    0.000
Residual Error    29       13199         455
Total             31      171437

Source         DF      Seq SS
PersInc         1      143610
NHMort          1       14628

Unusual Observations
Obs     PersInc      PersSav         Fit   StDev Fit   Residual   St Resid
 23        3802       142.00      184.11        4.42     -42.11     -2.02R
 24        4173       155.70      198.01        4.84     -42.31     -2.04R
 28        5264       272.60      217.14        6.92      55.46      2.75R
 30        5753       189.20      235.25        7.67     -46.05     -2.31R

R denotes an observation with a large standardized residual
```

Significance Testing and Goodness of Fit

In linear regression, we tested for a significant relationship by looking at the t- or F-ratios. In multiple regression, the two ratios test two different hypotheses. As before, the t-test is used to determine if a slope equals zero. Thus, in this case, we have two tests to perform:

For `PersInc`	**For `NHMort`**
$H_0: \beta_1 = 0$	$H_0: \beta_2 = 0$
$H_A: \beta_1 \neq 0$	$H_A: \beta_2 \neq 0$

The t-ratio and p-value in each row of the table of coefficients tell us whether or not to reject each of the null hypotheses. In this instance, at the .05 level of significance, we reject in both cases, due to the very low p-values. That is to say, both `PersInc` and `NHMort` have statistically significant relationships to `PersSav`.

The F-ratio in a multiple regression is used to test the null hypothesis that all of the slopes are equal to zero:

$$H_0: \beta_1 = \beta_2 = 0 \quad \text{vs. } H_A: H_0 \text{ is not true.}$$

Note that the alternative hypothesis is different from saying that all of the slopes are non-zero. If one slope were zero and the other were not, we would reject the null in the F-test. In the two t-tests, we would reject the null in one, but fail to reject it in the other.

Finally, let's consult r^2, the Coefficient of Multiple Determination. In the prior labs, we noted that the output reports both r^2 and "adjusted" r^2. It turns out that adding any X variable to a regression model will tend to inflate r^2. To compensate for that inflation, we adjust the coefficient of determination to account for both the number of x variables in the model, and for the sample size.[1] When working with multiple regression analysis, we generally want to consult the adjusted figure. In this instance, the addition of another variable really does help to explain the variation in Y.

In this regression, the adjusted r^2 equals 91.8%; in the prior session, using only income as the predictor variable, the adjusted r^2 was only 83%. We would say that, by including the mortgage rate in the equation, we are accounting for an additional 8.6% (91.8 - 83.2) of the total variation in aggregate personal savings.

Prediction and Residual Analysis

As in simple regression, we want to evaluate our models in terms of their predictive performance and the degree to which they conform to the assumptions concerning the random disturbance terms. Let us re-run the regression, making an estimate and saving the residuals for further analysis. Let's see how well the model performs in estimating savings for the year 1996, when US aggregate income was 6457.4 billion dollars and new home mortgage rates averaged 8.03 per cent.

---

[1] See your textbook for the formula for adjusted r^2. Note the presence of n and k (the number of predictors) in the adjustment.

Stat ➤ Regression ➤ Regression... We'll leave our variable selection as before, but in the **Options** dialog, enter the values `6457.4` and `8.03` into the **Prediction intervals for new observations** box. Also: in the **Storage** dialog, check **Residuals and Fits**.

> NOTE: In the **Prediction intervals for new observations** box of the **Options** dialog, you must enter the predictor values in the same order as the Predictor variables in the Regression dialog. That's how Minitab 'knows' that 6457.4 is the value for Income, not NHMort.

The regression output is virtually the same as before (it's the same regression), but now the prediction for 1996 appears at the end of the output. ***What is the predicted, or fitted, value for 1996? The actual figure for aggregate savings in 1996 was 265.5 billion dollars. Did the model come close? As before, note that the output includes both a 95% confidence interval and a 95% prediction interval. What do these intervals tell you?***

In this regression procedure, we saved the fitted and residual values. Let's now run a residual plot to validate the regression assumptions for this particular model.

Stat ➤ Regression ➤ Residual Plots... Select the residual and fitted values as appropriate.

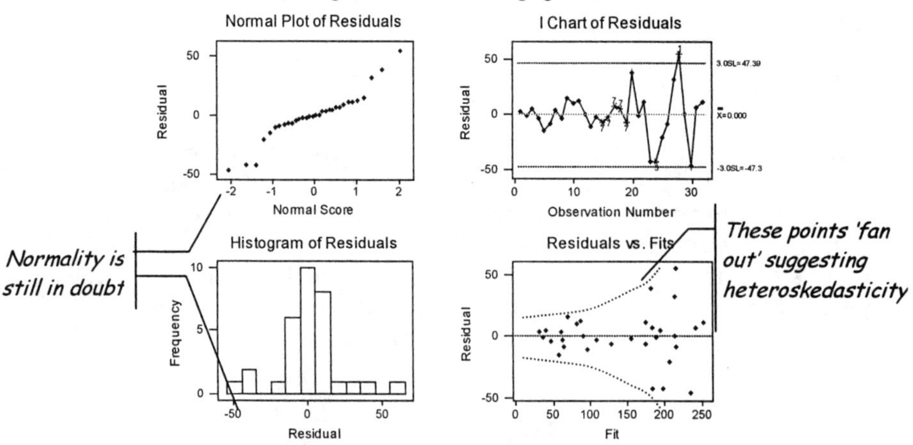

We interpret these residual plots exactly as we did before. In these particular graphs, you should note that there is doubt about both the normality condition and homoskedasticity. Besides explaining more variation than a simple regression model, a multiple regression model can sometimes resolve a violation of one of the regression assumptions. This is because the simple model may assign too much of the unexplained variation to ε, when it should be attributed to another variable. As such, the residuals are not genuinely random.

Adding More Variables

This model improved the simple regression. Let's see if we can improve this one further by adding another variable. Suppose we hypothesize that during times of high unemployment, people are reluctant to spend too much of their income, and therefore, have an extraordinary incentive to save, after controlling for income and mortgage rates. In other words, we want to add the Unemployment Rate variable (`Unemprt`) to the equation. Moreover, if this theory is right, we'd expect the coefficient of `Unemprt` to be positive.

🖱 **Stat ➤ Regression ➤ Regression...** Add `Unemprt` to the list of predictors. Before clicking **OK**, click on **Options** and delete the predictor values. Then perform the regression.

Compare this regression output to the earlier results. ***What was the effect of adding this new variable to the model?*** Obviously, we have an additional coefficient and t-ratio. ***Does that t-ratio indicate that*** `Unemprt` ***has a significant relationship to aggregate savings, when we control for income and interest rates?***

What else changed? Look, in particular, at the adjusted r^2, the ANOVA, and the values of the previously estimated coefficients. Can you explain the differences you see?

The addition of a new variable can also have an impact on the residuals. In general, each new model will have a new residual plot. The fits and residuals for this model are in the last two columns of the worksheet, designated FITS2 and RESI2. ***Examine this set of residual graphs, and see what you think. Do the least squares assumptions appear to be satisfied?***

A New Concern

In a model including income, mortgage interest and unemployment rates, we have accounted for about 93% of the variation in Aggregate Personal Savings, and have a standard error of the regression of about 20 billion dollars. This is quite an improvement over the simple model in which r^2 was 83% and *s* was 30 billion dollars. Suppose we wanted to try to improve the model even further, and hypothesize that at times of high inflation, people might save even more than usual, with consumption unattractive due to high prices. Let's add the Consumer Price Index (CPI) to the model, and re-run the regression with CPI added to the list of predictors.

Look at this regression output, and take special note of the total variation explained, the standard error (*s*), and the other regression coefficients and test statistics. *None* of the estimated slopes appear to be statistically significant at the .05 level! What is happening here?

This is an illustration of a special concern in multiple regression: **multicollinearity**. When two or more of the predictor variables are highly correlated in the sample, the regression procedure cannot determine which of the predictor variables concerned is associated with changes in Y. In a real sense, regression cannot "disentangle" the individual effects of each X. In this instance, the culprits are CPI and Personal Income.

Stat ➤ Basic Statistics ➤ Correlation... Select all five variables for this correlation matrix: **PersSav, PersInc, NHMort, Unemprt, CPI**

Correlations (Pearson)

```
          PersSav   PersInc    NHMort   Unemprt
PersInc    0.915
           0.000

NHMort     0.464     0.194
           0.007     0.287

Unemprt    0.527     0.285     0.731
           0.002     0.114     0.000

CPI        0.946     0.990     0.318     0.388
           0.000     0.000     0.076     0.028
```

Note that CPI and Savings are highly correlated (.946) which ordinarily is good; however, also note that CPI and Personal Income have a nearly perfect correlation of .990. This is the root of the problem, and with this sample, can only be resolved by eliminating one of the two variables from the model. Which should we eliminate? We should be guided both by theoretical and numerical concerns: we have a very strong theoretical reason to believe that income belongs in the model, but CPI has a stronger numerical association with savings. Given a strong theoretical case, it is probably wiser to retain income and omit the CPI.[2]

Another Example

Let's see another example, this time including a qualitative variable in the analysis. Open the worksheet called **Colleges**. Our concern in this problem is what admissions officers call "Admissions Yield." When you were a high school senior, your college sent acceptance letters out to many students. Of that number, many chose to attend a different school. The "yield" refers to the proportion of students who actually enroll, compared to the number admitted. In this regression model, we will concentrate on the relationship between the number of seniors a college accepts, and the number who eventually enroll. Let's look at that relationship graphically.

- Construct a scatterplot of `NewEnrol` (Y) vs. `AppAcc` (X)—the number of freshmen enrolling in each school versus the number of applications accepted. You have seen this plot in an earlier session.

Since there are nearly 1,300 schools in this sample, the graph is very dense. Notice that the points are tightly clustered in the lower left of the graph, but fan out as you move to the upper right. Even in this scatterplot, you can see evidence of heteroskedasticity. ***Why do you think that might occur? That is, what would cause the heteroskedasticity?***

- Now let's run the regression and evaluate the model. Run a regression using `NewEnrol` as the response and `AppAcc` as the sole predictor. For now, also click on **Storage**, and *uncheck* **Fits** and **Residuals**; we won't save them this time. As an

[2] We could keep income, and use the annual *change* in the CPI, which measures inflation in the given year. The idea here is to illustrate the effects of multicollinearity, and to emphasize that the problem should not be ignored.

alternative way of seeing residual graphs, click on the **Graphs** button, and select **Histogram of residuals** and **Residuals vs. Fits**. Click OK, and then run the regression.

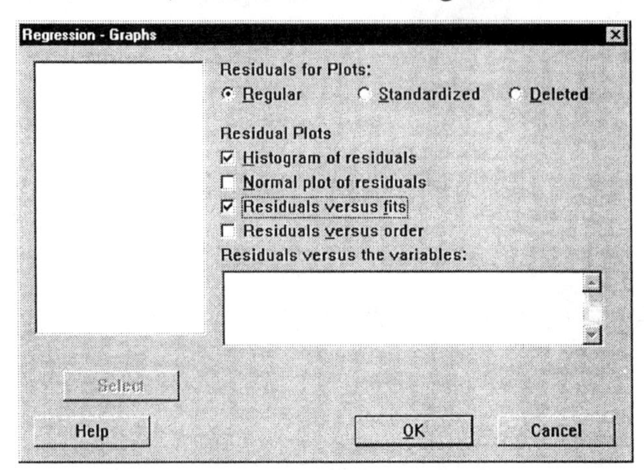

In addition to the regression output, you'll see two graphs, as selected in the dialog. What do they suggest about the validity of the least squares assumptions in this case? These residuals form a generally normal histogram, but do seem to fan out from left to right in the residuals versus fits plot (as we predicted from the scatterplot).

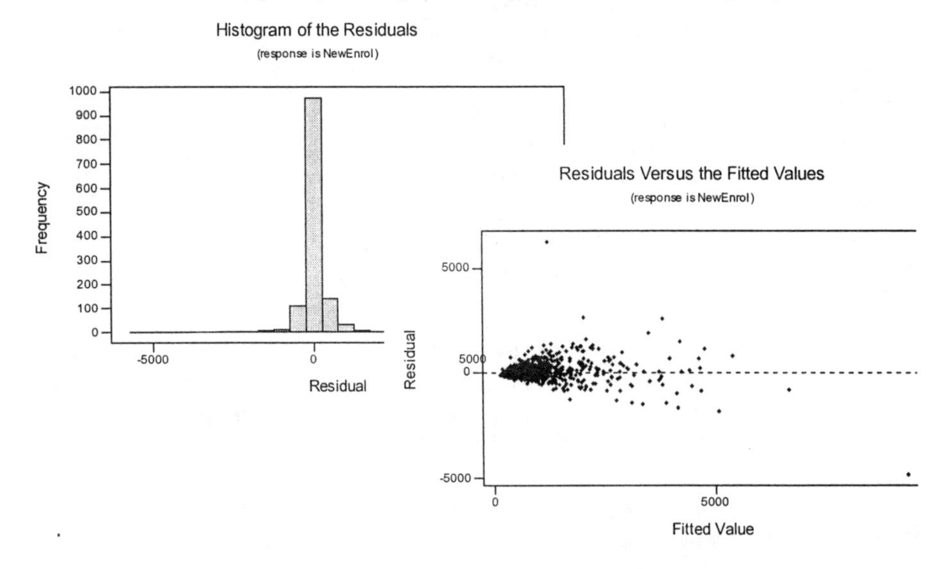

Like most regressions, this one has strengths and flaws. Examine the regression output, including the residual graphs, and evaluate the model, considering these questions:

- ***Does the coefficient of*** `AppAcc` ***have the expected sign?***
- ***Is the relationship statistically significant?***
- ***How good is the fit?***
- ***Are there many unusual observations?***
- ***Are the residuals normally distributed?***
- ***Are the residuals homoskedastic?***

Working with Qualitative Variables

In this regression, the number of applications accepted accounts for a large percentage of the variation in enrollments. Another potentially important factor in explaining the differences in enrollment, and in admissions yield in general, is whether or not a school is publicly funded.

Whether a school is public or private is, of course, a categorical variable. All of the variables we have used in regression analysis so far are quantitative. It is possible to include a qualitative predictor variable in a regression analysis if we come up with a way to represent a categorical variable numerically.

We do this with a fairly simple "trick" known as a **dummy variable.** A dummy variable is an artificial binary variable which assumes arbitrarily chosen values to represent two distinct categories.[3] Ordinarily, we set up a new variable which equals 0 for one category, and 1 for the other. In this dataset, we have a variable called **`PubPvt`** which happens to equal 1 for public colleges and 2 for private colleges. We can use that variable in this case.

Let's re-run the regression, simply adding **`PubPvt`** to the **Predictors** box.

Now look at the regression results in the Session Window, and the regression equation itself. Think for a moment about the meaning of the coefficient of the **`PubPvt`** variable. For a public college or university, **`PubPvt`** = 1, and this equation becomes:

```
NewEnroll = 777 + 0.316 AppsAcc - 360(1)
          = 417 + 0.316 AppsAcc
```

[3] It is possible to represent multinomial variables using several dummy variables. Consult your textbook or instructor.

On the other hand, for a private school, the equation is:

```
NewEnroll = 777 + 0.316 AppsAcc - 360(2)
          =  57 + 0.316 AppsAcc
```

In other words, we are looking at two parallel lines whose intercepts differ by 360 students. The impact of the dummy variable, introduced into the equation in this way, is to alter the *intercept* for the two different categories.

Now that we know what the estimated equation is, let's go on to evaluate this particular model, as we did earlier.

- ***Do the estimated coefficients have the expected signs?***
- ***Are the relationships statistically significant?***
- ***How good is the fit?***
- ***Are there many unusual observations?***
- ***Are the residuals normally distributed?***
- ***Are the residuals homoskedastic?***

Moving On...

Now apply the techniques learned in this session to the following questions.

Colleges

1. We've found two variables which help to estimate new enrollment. Let's see if we can expand the model to do a more complete job. Your task is to choose one more variable from among the following, and add it to the regression model:

   - `Tuit_out`
   - `Top10`
   - `FacTerm`

   You may choose any one you wish, as long as you can explain how it logically might affect new enrollments, once acceptances and public/private are accounted for. Then run the regression model including the new variable, and evaluate the regression in comparison to the one we have just done. In your evaluation, you should comment on these questions:

   - Are signs of coefficients correct?

- Does the model explain substantially more of the variation in y than the prior model?
- Are the relationships statistically significant?
- Do the residuals suggest that the assumptions are satisfied?
- Is there any evidence of a problem with multicollinearity?

Bodyfat

2. Develop a multiple regression model to estimate the body fat percentage of an adult male (`FatPerc`), based on one or more of the following easily-measured quantities:
   - Age (years)
   - Weight (pounds)
   - Abdomen circumference (in cm)
   - Chest circumference (in cm)
   - Thigh circumference (in cm)
   - Wrist circumference (in cm)

   You should refer to a matrix plot and/or correlation matrix to help select variables. Your model may contain any or all of the listed predictor variables. The model which you select must (a) make sense, (b) have good significance test results, (c) have acceptable residuals (d) have as high an r^2 as you can get, and (e) have no multicollinearity. Also, discuss possible logical problems with using a linear model to estimate body fat percentage.

Sleep

3. Develop a multiple regression model to estimate the total amount of sleep (`Sleep`) required by a mammal species, based on one or more of the following variables:
   - Body weight
   - Brain weight
   - Lifespan
   - Gestation

   You should refer to a matrix plot and/or correlation matrix to help select variables. Your model may contain any or all of the listed predictor variables. The model which you select must (a) make sense, (b) have good significance test results, (c) have acceptable residuals (d) have as high an r^2 as you can get, and (e) have no multicollinearity. Also, discuss possible logical problems with using a linear model to estimate sleep requirements.

Labor

4. Develop a multiple regression model to estimate the mean number of new weekly unemployment insurance claims. [NOTE: The variable name is A-zero-M-zero-zero-five (`A0M005`).] Select variables from this list, based on theory, and the impact of each variable in the model.
   - `A0M441` Civilian labor force
   - `A0M060` Ratio of index of newspaper ads to number unemployed
   - `A0M037` Number of people unemployed
   - `A0M043` Civilian unemployment rate

   You should refer to a matrix plot and/or correlation matrix to help select variables. Your model may contain one, two, three, or all four predictor variables, aiming for a model which (a) makes sense, (b) has good significance test results, (c) has acceptable residuals (d) has as high an r^2 as you can get, and (e) has no multicollinearity.

Utility

5. During the study period, we added a room to this house, and thereby increased its heating needs. Use the variable `NewRoom` in a multiple regression analysis (including measures of temperature) to estimate the additional number of therms per day consumed as a consequence of enlarging the house. As in prior problems, comment on the significance tests and Goodness of Fit measures, as well as the residual analysis.

Session 18

Non-Linear Models

Objectives

In this session, you will learn to do the following:

- Improve a regression model by transforming the original data
- Interpret coefficients and estimates using transformed data

When Relationships Are Not Linear

In our regression models thus far, we have assumed *linearity*; that is, that Y changes by a fixed amount whenever an X changes by one unit, other things being equal. The linear model is a good approximation in a great many cases. However, we also know that some relationships probably are not linear. Consider the "law of diminishing returns" as illustrated by a weight-loss diet. At first, as you reduce your calories, pounds may fall off quickly. As your weight goes down, though, the rate at which the pounds fall off may diminish.

In such a case, X and Y (calories and weight loss) are indeed related, but *not in a linear fashion*. This session provides some techniques that we can use to fit a *curve* to a set of points. The basic strategy in each case is the same. We will attempt to find a function whose characteristic shape approximates the curve of the points. Then, we'll apply that function to one or more of the variables in our worksheet, until we have two variables with a generally linear relationship. Finally, we'll perform a linear regression using the transformed data. We will begin by using an artificial example.

A Simple Example

Let's begin with a very familiar non-linear relationship between two variables, in which Y varies with the square of X. The formal model (known as a quadratic model) might look like this:

$$Y = 3X^2 + 7 + \varepsilon$$

In fact, let's create an artificial set of data that reflects an exact relationship.

- Label the first three columns of a blank Minitab worksheet as `X`, `Xsqr`, and `Y`.

- **Calc ➤ Make Patterned Data ➤ Simple Set of Numbers...** Indicate that you want to store the data in `C1` (`X`), that the first value is `1`, and the last value is `20`. Then click **OK**. This will fill the X column with the values 1 through 20.

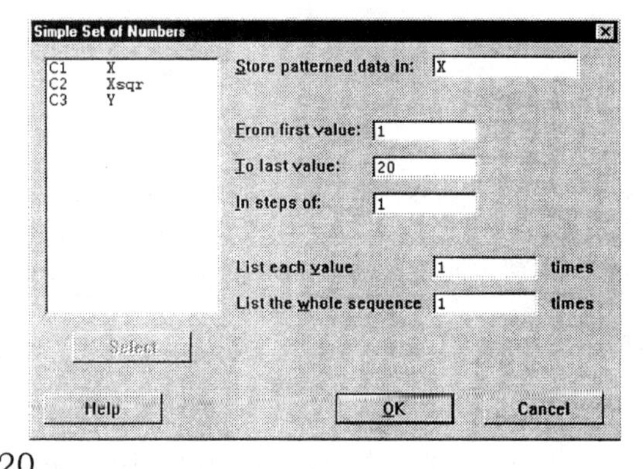

- **Calc ➤ Calculator**. In `Xsqr`, calculate the square of X.

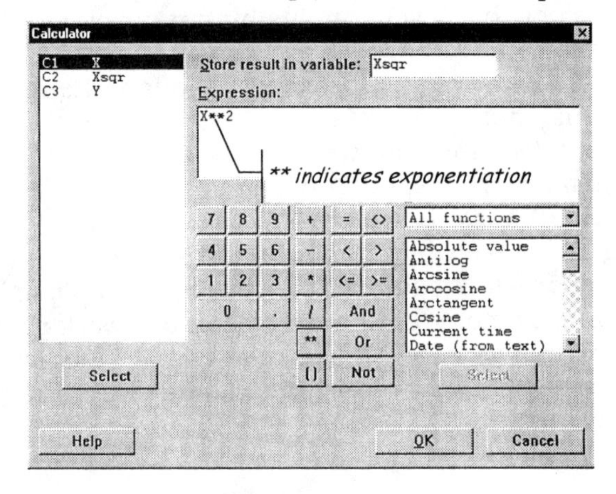

- Now, set Y equal to $3*X^2 + 7$. Bring up the calculator again, storing the result in `Y`. The expression should read:

```
3*Xsqr + 7
```

- Click **OK** on the calculator, and look at the Data Window. You will see three columns of numbers. The first row is 1, 1, 10 and the second row is 2, 4, 19 (see first few lines below).

	C1	C2	C3	C4
↓	X	Xsqr	Y	
1	1	1	10	
2	2	4	19	
3	3	9	34	
4	4	16	55	
5	5	25	82	
6	6	36	115	
7	7	49	154	
8	8	64	199	

If we plot Y versus X, and then plot Y versus Xsqr, the graphs look like this:

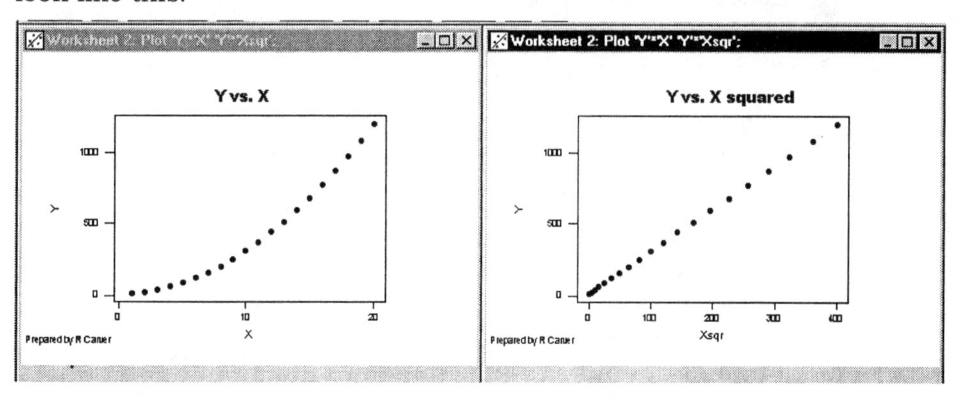

In the left graph, we see the distinct parabolic shape of a quadratic function. Y increases each time that X increases, but does so by an *increasing* amount. Clearly, X and Y are related, but just as clearly, the relationship is **curvilinear**. In the second graph, we have a perfectly straight line that is an excellent candidate for simple linear regression. ***If you were to run a regression of Y on Xsqr, what would the slope and intercept be?*** *(Do it, and check yourself.)*

This illustrates the strategy we noted above: When we have a curved relationship, we'll try to find a way to transform one or more of our variables until we have a graph which looks linear. Then we can apply our familiar and powerful tool of linear regression to the *transformed* variables. *So long as we can transform one or more variable and retain the basic functional form of y as a sum of coefficients times variables*, we can use linear regression to fit a curve. That is the basic idea underlying the next few examples.

Some Common Transformations

In our first artificial example, we squared a single explanatory variable. As you may recall from your algebra and calculus courses, there are many common curvilinear functions, such as cubics, logarithms, and exponentials. In this lab, we'll use a few of the many possible transformations, just to get an idea of how one might create a mathematical model of a real world relationship.[1]

Let's begin with an example from Session 15 (and about 400 years ago): Galileo's experiments with rolling balls. Recall that the first set of data plotted out a distinct curve:

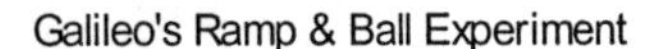

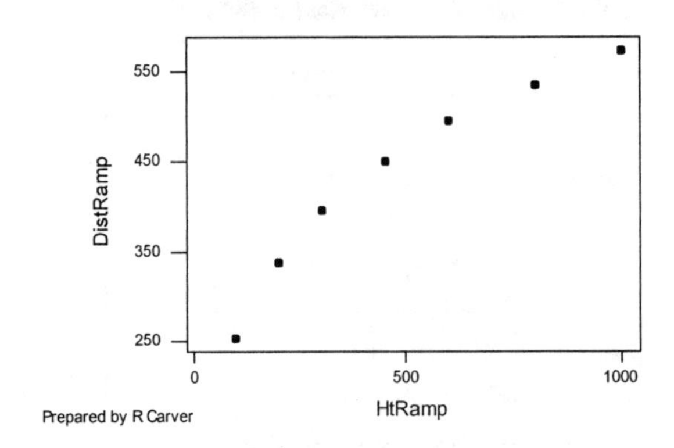

Note that the ball rolled further when started from a greater height, but that the increased horizontal roll diminishes as heights increase. A straight line is not a bad fit, but we can do better with a

[1] Selection of an appropriate model should be theory-driven, but sometimes can involve trial and error. The discussion of theoretical considerations is beyond the scope of this manual; consult your primary text and instructor for more information.

different functional form. In fact, Galileo puzzled over this problem for quite some time, until he eventually reasoned that horizontal distance might vary with the square root of height.[2] If you visualize the graph of $y = +\sqrt{x}$, it looks a good deal like the scatterplot below: it rises quickly from left to right, gradually becoming flatter.

Open the worksheet called **Galileo**, and use the **Calculator** to create a new variable equal to the *square root* of `HtRamp`. Label that variable `SqrtHt`.

Graph ➤ Plot `DistRamp` is on the Y axis, and `SqrtHt` on the X axis, generating this graph:

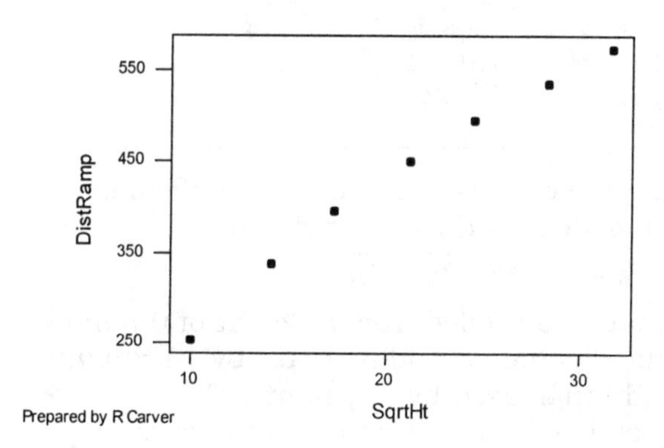

While not as dramatic as our first dataset, we see that this transformation does tend to straighten out the curved line. What is more, it makes theoretical sense that friction will tend to "dampen" the horizontal motion of a ball rolled from differing heights, just as the square root function represents. Other functional transformations may align the points more completely, but don't make as much sense, as we'll see later.

---

[2] For an interesting account of his work on these experiments, see Dickey, David A. and Arnold, J. Tim, "Teaching Statistics with Data of Historic Significance." *Journal of Statistics Education,* v. 3, no. 1, 1995.

Now that we've straightened the dots, what's next? First, we run a regression, with **DistRamp** as the Response variable, and **SqrtHt** as the lone Predictor. Use the regression command to generate these results:

Regression Analysis

```
The regression equation is
DistRamp = 129 + 14.5 SqrtHt

Predictor        Coef      StDev          T        P
Constant       129.02      18.37       7.02    0.001
SqrtHt        14.5150     0.8274      17.54    0.000

S = 15.69       R-Sq = 98.4%     R-Sq(adj) = 98.1%

Analysis of Variance

Source            DF         SS         MS        F        P
Regression         1      75791      75791   307.72    0.000
Residual Error     5       1231        246
Total              6      77022
```

Let's focus on interpreting the coefficients and the significance tests for starters. Our estimated model could be re-written as:

$$Dist = 129.02 + 14.515\sqrt{RampHt}$$

The intercept suggests that a ball rolled from a height of 0 *punti* would roll 129 *punti*, and that the distance would increase by 14.5 *punti* each time the square root of height increased by one *punto*. The significance tests strongly suggest a statistically significant relationship between the two variables and the fit is excellent (r^2-adj. = 98.1%).

Using the equation, we can estimate the distance of travel by simply substituting a height into the model. For instance, if the initial ramp height were 900 *punti*, we would have:

$$\begin{aligned} Dist &= 129.02 + 14.515\sqrt{900} \\ &= 129.02 + 14.515(30) \\ &= 564.47 \textit{ punti} \end{aligned}$$

Note that we must take care to transform our x-value here in order to compute the estimated value of y. Our result is calculated using the square root of 900, or 30.

Like any regression analysis, we must also check the validity of our assumptions about ε. The next example includes that analysis, as well as another curvilinear function.

Another Quadratic Model

Non-linear relationships crop up in many fields of study. Let's return to the relationship between aggregate personal savings and aggregate personal income, which we first saw in our very first linear regression lab. Open the **Us** worksheet.

As you may recall, the least squares line did not fit these points well, and there were several problems in the residuals. As a starting point, let's look at the simple linear model once more.

Stat ➤ Regression ➤ Fitted Line Plot... We'll use this command to run the regression, since it allows us to use quadratic and cubic transformations very easily. In this dialog, specify the two variables as shown, and click **Storage.** Check **Residuals** and **Fits** in the storage dialog, and click **OK** in both dialogs.

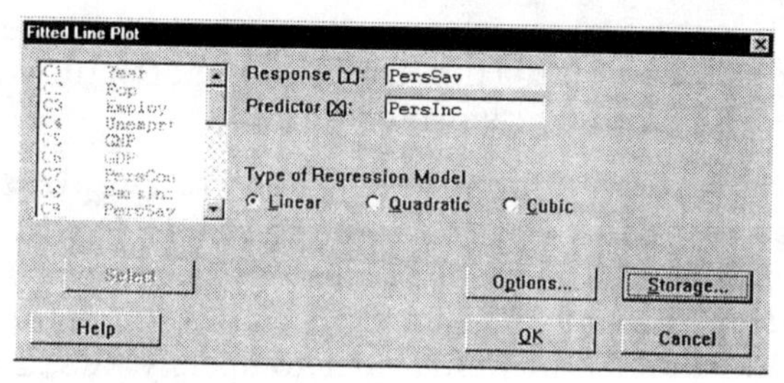

As you can see in the resulting graph, the points arc around the fitted line, and r^2 is only .84. In the Session Window, note that the estimated slope of the line is statistically significant.

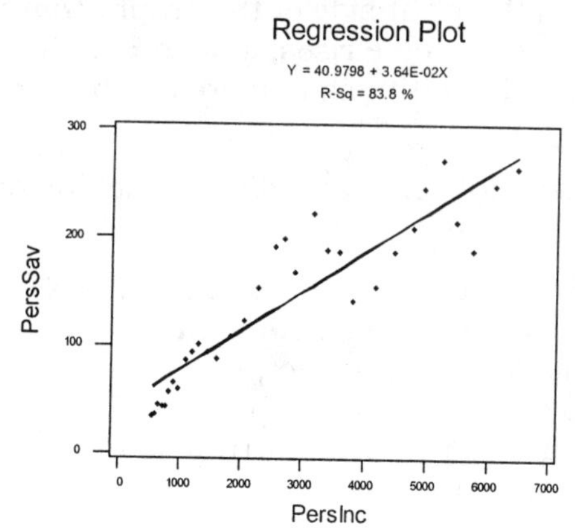

Now, we should examine the residuals.

Stat ➤ Regression ➤ Residual Plots Your residuals are in RESI1 and the fits are FITS1. ***What deficiencies do you find in these residuals? Do they suggest normally distributed disturbances? Are the disturbances independent of one another? Does the linearity assumption seem reasonable?***

Look back at the fitted line plot. Can you visualize the left half of an inverted parabola? Such a pattern might fit this model:

$$Savings_i = \beta_0 + \beta_1 Income_i + \beta_2 Income_i^2 + \varepsilon_i$$

Return to the Fitted Plot command dialog again, but this time click on the option button labeled **Quadratic.** You will see another graph, get new regression output, and a new set of residuals and fitted values. This, then, is a short-cut way to run the regression with transformed data.

How does this regression equation compare to the first one? What do the two slopes tell you? Are the results statistically significant? Has the standard error decreased? Has the coefficient of multiple determination improved? What about the earlier problems with the residuals? (Hint: Remember to specify RESI2 and FITS2 this time.) ***What strengths and weaknesses do you find in these residuals?***

One theoretical problem with this quadratic model is that it begins to slope down as you move to the right side of the graph; that is, it suggests that savings will decline when income rises, say, to $7,000 billion. Let's try the cubic model, which continues to increase throughout the values of income.

Once more, edit the **Fitted Plot** command dialog, clicking on the option button labeled **Cubic**.

How does the cubic model compare to the quadratic? Are the residuals better in this model or in the quadratic model?

As you can see, the cubic model is not perfect, but from this short example, you can get a sense of how the technique of data transformation can become a very useful tool. Let's turn to another example, using yet another transformation.

A Logarithmic Transformation

For our last illustration, we'll return to the household utility dataset (**Utility**). As in prior labs, we'll focus on the relationship between gas consumption (`GaspDay`) and mean monthly temperature (`MeanTemp`).

You may recall that there was a strong negative linear relationship between these two variables when we performed a simple linear regression (r^2 was .864). There were some problems with the regression, though. The plot of Residuals vs. Fitted values suggested some non-linearity.

When you think about it, it makes sense that the relationship can't be linear over all possible temperature values. In the sample, as the temperature rises, gas consumption falls. In the linear model, there must be some temperature at which consumption would be negative, which obviously cannot occur. A better model would be one in which consumption falls with rising temperatures, but then levels off at some point, forming a pattern similar to a natural logarithm function. The natural log of temperature serves as a helpful transformation in this case; that is, we will perform a regression with `GaspDay` as Y, and *ln*(`MeanTemp`) as X.

Calc ➤ Calculator Type `LnTemp` in the box marked **Store result in**, and then type `LOGE(MeanTemp)` in the expression box. This indicates that you want to create a new variable, equal to the natural (or base e) log of temperature. Click **OK**.

Make two scatterplots: `GaspDay` vs. `MeanTemp` and `GaspDay` vs. `LnTemp`. The second graph is more linear than the first.

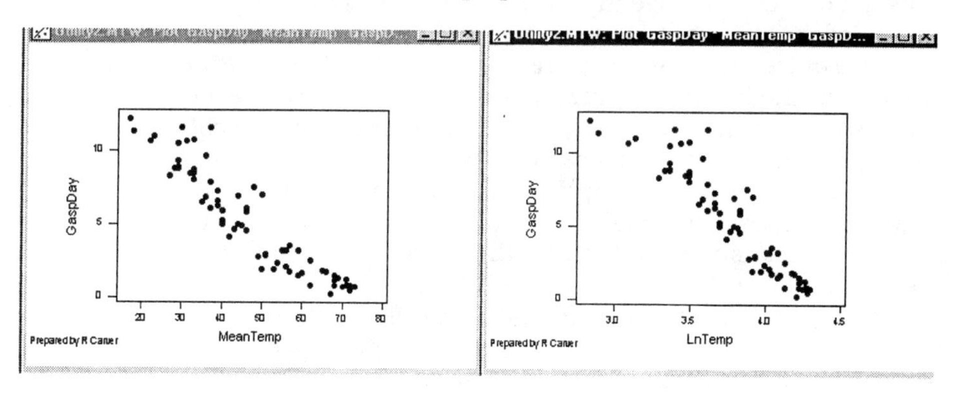

Stat ➤ Regression ➤ Regression... The response variable is `GaspDay`, the predictor is `LnTemp`. Save the fits and residuals.

Regression Analysis

```
The regression equation is
GaspDay = 40.6 - 9.34 LnTemp

71 cases used 10 cases contain missing values

Predictor        Coef       StDev          T        P
Constant       40.597       1.704      23.83    0.000
LnTemp        -9.3380      0.4489     -20.80    0.000

S = 1.322        R-Sq = 86.2%     R-Sq(adj) = 86.0%

Analysis of Variance

Source           DF          SS          MS        F         P
Regression        1      755.71      755.71   432.72     0.000
Residual Error   69      120.50        1.75
Total            70      876.21
```

What are the strengths and weaknesses of this regression? What is your interpretation of the residual analysis? Check back through your earlier notes, and compare this regression to the simple regression involving `GaspDay` and `MeanTemp`. In our earlier model, a one-degree increase in temperature was associated with a decrease of 0.22 therms. ***What does the slope in the new model mean?***

Apply the same logic as we always have. The slope is the marginal change in Y, given a one unit change in X. Since X is the natural log of temperature, the slope means that consumption will decrease 9.338 therms when the *log* of temperature increases by 1. The key here is that one-unit differences in the natural log function are associated with ever-increasing changes in temperature as we move up the temperature scale.

As you look at your notes, you'll see that we once used this regression model to predict gas consumption for a month in which temperature was 40 degrees. Suppose we want to do that again with the transformed data. We can't simply substitute the value of 40, since X is no longer temperature, but rather it is the natural logarithm of temperature. As such, we must substitute $ln(40)$ into the estimated equation. Doing so will yield an estimate of gas consumption. ***What is the estimated consumption for a month with a mean temperature of 40 degrees?***

In the simple linear model, we obtained a negative consumption estimate for a mean temperature of 75°. ***Estimate consumption with this new model, using the ln(75). Is this result negative?***

Adding More Variables

We are not restricted to using simple regression or to using a single transformed variable. All of the techniques and caveats of multiple regression still apply. In other words, one can build a multiple regression model that involves some transformed data and other variables as well. In addition, we can transform the response variable. This requires additional care in the interpretation of estimates, because the fitted values must be "un-transformed" before we can work with them.

Moving On...

Galileo

1. Return to the data from the first Galileo experiment (first two columns), and using the Fitted Line Plot command, fit quadratic and cubic models. Discuss the comparison of the results.
2. Use the two new models to estimate the horizontal roll when a ball is dropped from 1,500 *punti*. Compare the two estimates to an estimate using the square root model. Comment on the differences among the three estimates, select the estimate you think is best, and explain why you chose it.
3. Fit a curvilinear model to the data in the third and fourth column. Use both logic and statistical criteria to select the best model you can formulate.

Bodyfat

4. Refer to question 2 in Session 17. Using one or more of the variables listed there with an appropriate transformation, re-estimate a model to estimate the body fat percentage of an adult male. Follow the same criteria as in that question.

Sleep

5. Refer to question 3 in Session 17. Using one or more of the variables listed there with an appropriate transformation, re-estimate a model to estimate the total amount of sleep required by a mammal species. Follow the same criteria as in that question.

Labor

6. Refer to question 4 in Session 17. Using one or more of the variables listed there with an appropriate transformation, re-estimate a model to estimate the mean number of new weekly unemployment insurance claims (A0M005). Follow the same criteria as in that question.

Session 19

Basic Forecasting Techniques

Objectives

In this session, you will learn to do the following:

- Identify common patterns of variation in a time series
- Make and evaluate a forecast using Moving Averages
- Make and evaluate a forecast using Exponential Smoothing
- Make and evaluate a forecast using Trend Analysis

Detecting Patterns over Time

In the most recent lab sessions, our concern has been with building models that attempt to account for variation in a dependent, or response, variable. These models, in turn, can be used to estimate or predict future or unobserved values of the dependent variable.

In many instances, though, variables behave fairly predictably over time. In such cases, we can use techniques of *Time Series Forecasting* to predict what will happen next. There are many such techniques available; in this lab, we will experiment with only three of them. This session is very much an introduction to these tools. Consult your textbook for additional techniques.

Recall that a **time series** is a sample of repeated measurements of a single variable, observed at regular intervals over a period of time. The length of the intervals could be hourly, daily, monthly; what is most important is that it be *regular*. When we examine a time series, we typically expect to find one or more of the following common idealized patterns, often in combination with one another:

- ***Trend*** – a general upward or downward pattern over a long period of time, typically years. A time series showing no trend is sometimes called a *stationary time series.*
- ***Cyclical** variation* – a regular pattern of up-and-down waves, such that peaks and valleys occur at regular intervals. Cycles emerge over a long period of years.
- ***Seasonal** variation* – a pattern of ups and downs within a year, repeated in subsequent years. Most industries have some seasonal variation in sales, for example.
- ***Random, or irregular**, variation* – movements in the data, which cannot be classified as one of the preceding types of variation.

Let's begin with some real-world examples of these patterns. Be aware that the four time series components just listed are ideal categories. It is rare to find a real time series which is a "pure" case of just one component or another.

Some Illustrative Examples

Open the file **Us**. All of the variables in this file are measured annually. Therefore, we cannot find seasonal variation here. However, there are some good examples of trend and cycle, as well as irregular variation.

Graph ➤ Time Series Plot... In the dialog box, we'll select the variables `NHMort`, `Unemprt`, `Starts`, `M1`, and `Pop`. This will create five graphs.

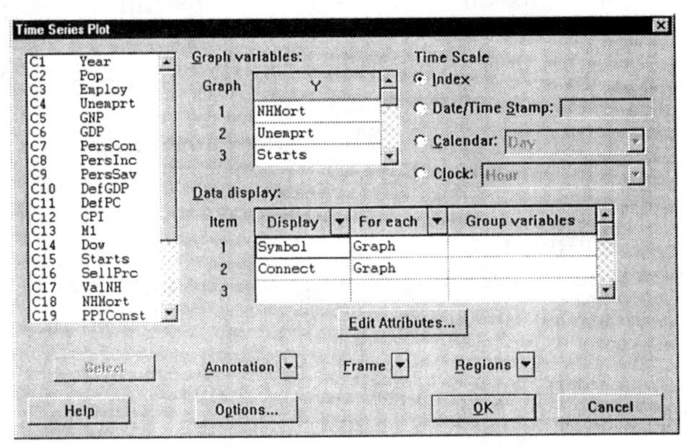

The five variables are these (1965–1996):

- **NHMort**: New Home Mortgage Interest Rate
- **Unemprt**: Unemployment Rate (%)
- **Starts**: Housing Starts (thousands), i.e., the number of new homes on which construction started in the year
- **M1**: Money Supply (billions of $)
- **Pop**: Population of the US (thousands).

The topmost graph shows the population of the United States, and one would be hard-pressed to find a better illustration of a linear trend. During the period of the time series, population has grown by a nearly constant number of people each year. It is easy to see how we might extrapolate this trend.

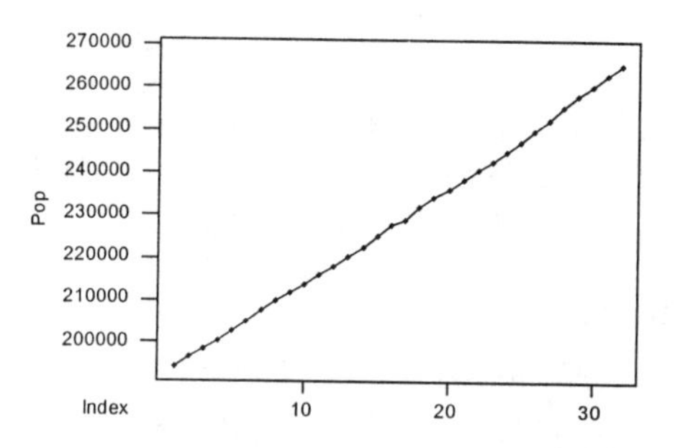

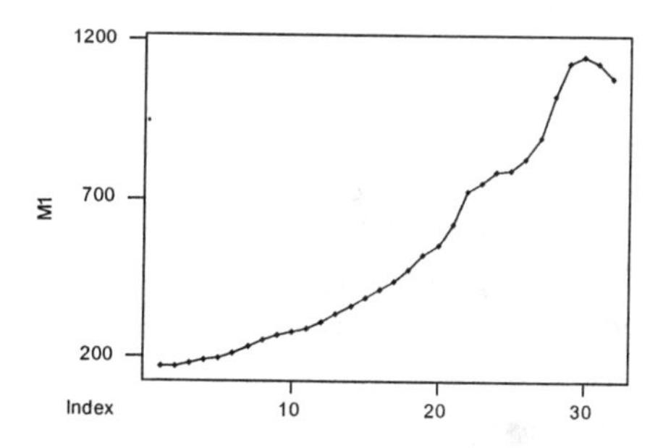

What do you see in the next graph (M1)? There is also a general trend here, but it is *non-linear*. If you completed the session about non-linear models, you might have some ideas about a functional form which could describe this curve. In fact, as we'll see later, this graph is a typical example of growth which occurs at a constant **percentage** rate, or exponential growth.

The third graph in the 'stack' (housing starts) is a rough illustration of **cyclical** variation, combined with a moderate negative trend. Although the number of starts increases and decreases, the general pattern is downward, with peaks and valleys spaced fairly evenly.

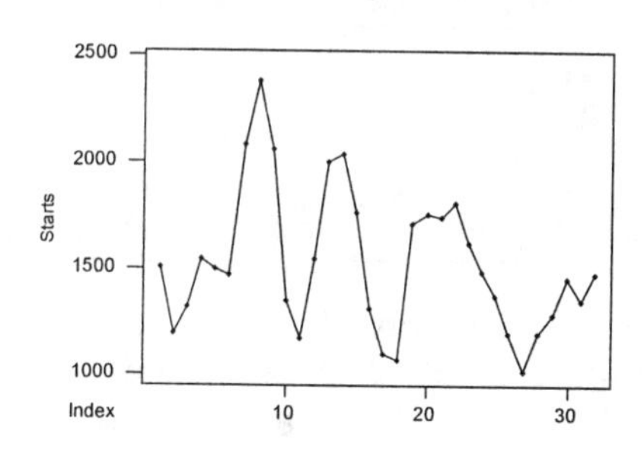

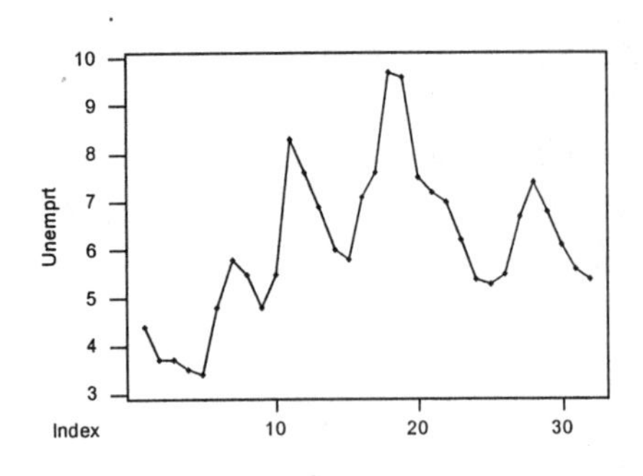

The fourth graph, showing the unemployment rate during the period, in contrast to the third graph has unevenly spaced peaks, and the upward trend visible on the left side of the graph appears to be flattening or even declining on the right side. The irregularities here suggest a sizable erratic component. This graph also illustrates the way in which various patterns might combine in a graph.

Finally, the graph of mortgage rates is almost entirely irregular movement. The pattern that we see is not easily classified as one of the principal components noted earlier.

To see seasonal variation, we return to the **Utility** file, with home heating data from New England.

Open the worksheet **Utility**. We'll focus on the variable `MeanTemp`, which represents the mean temperature for the month. It certainly makes sense to expect that we'll see seasonal variation.

Graph ➤ Time Series Plot... This time, to help us see the seasonal periodicity, we'll add some lines to the graph. Select the variable `MeanTemp`, and then click on the word **Frame** in the dialog box. Select **Grid**... from the list. This brings up the dialog shown below. Enter `X1` in the **Direction** box, and click **OK**. Click **OK** in the main Time Series Plot dialog as well.

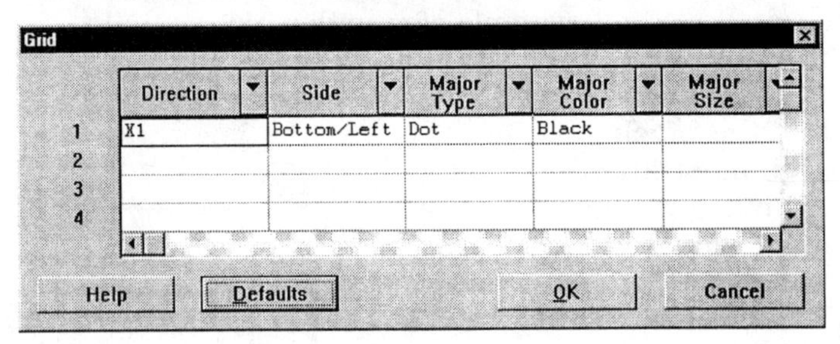

Look at the graph. Note that the first "valley" occurs at the fifth observation (January, 1991). Now scan across to twelve months later, at observation 17. Do the same all the way across the graph. Do you have any questions?

We can exploit patterns such as these to make predictions about what will happen in future observations. Any such attempt to predict the future will of necessity yield imperfect forecasts. Since we cannot eliminate errors in our predictions, the trick in forecasting is to *minimize* error. Each of the techniques illustrated below is suited to modeling different time series patterns, and each has its virtues. Each also inevitably involves *forecast errors,* which are the differences between a forecast value and the (eventual) observed value.

The first two techniques we'll explore are most useful for "smoothing out" the erratic jumps of irregular movements. They are known as Moving Averages and Exponential Smoothing. For each of the techniques which follows, we'll illustrate the technique using a single time series, the dollar value (in millions) of exports of domestic agricultural products from January, 1948 through March, 1996. That time series is found in the file called **Eximport**. Open that file.

Forecasting Using Moving Averages[1]

Moving Averages is an appealing technique largely due to its simplicity. We generate a forecast by merely finding the simple average of a few recent observations. The key analytical question is to determine an appropriate number of recent values to average. In general, we make that determination by trial and error, seeking an averaging period which would have minimized forecast errors in the past.

Thus, in a Moving Average analysis, we select a moving average 'length' or interval, compute retrospective "forecasts" for an existing set of data, and then compare the forecast figures to the actual figures. We summarize the forecast errors using some standard statistics explained below. We then repeat the process several more times with different moving average intervals to determine which interval length would have performed most accurately in the past.

Before making any forecasts at all, let's look at the time series.

- Construct a time-series plot of A0M604. ***Comment on any patterns or unusual features of this plot.***

- **Stat ➤ Time Series ➤ Moving Average...** In the dialog, select the variable A0M604. Specify a **MA Length** of 6 months, and

[1] In the remainder of the session, we will assume that your primary text covers the theory and formulas involved in these techniques. Consult your text for questions about these issues.

indicate that you wish to generate 1 forecast (for April, 1996). Finally, title this appropriately to indicate that it is a six-month moving average.

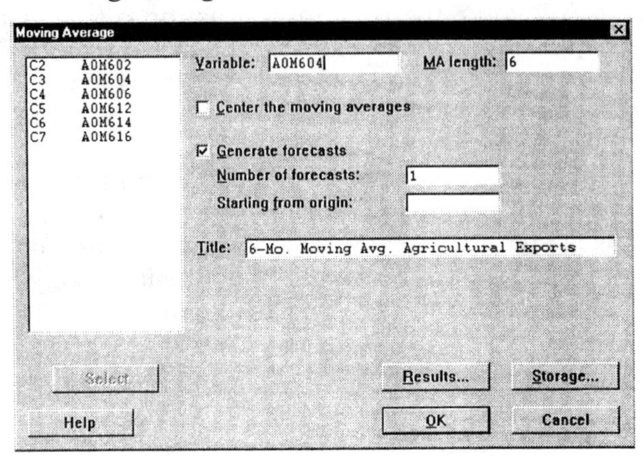

Maximize the resulting graph window. Due to the large sample, the graph is crowded, but you can distinguish between the observed values (black) and the smoothed six-month moving average series (blue). Note that the blue line 'shadows' the black, slightly to the right. Note also that the blue line is slightly smoother than the black, never quite reaching the extreme values of the black dots. At the far right is our one forecast, in red. Checking the Session Window, we find that the estimate for April, 1996 is for $4,851 million in agricultural exports.

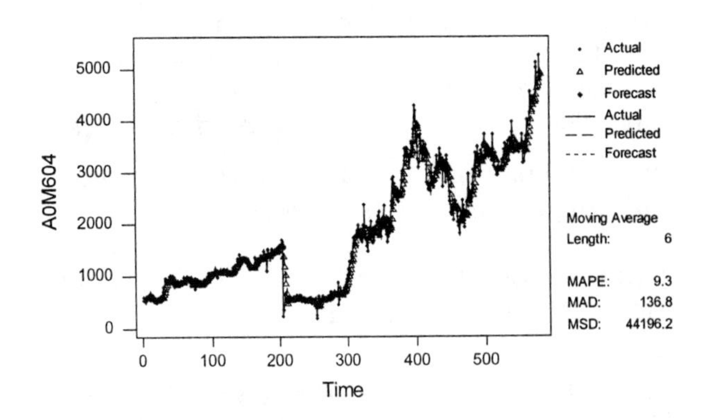

```
Moving average

Data        A0M604
Length      615.000
NMissing    36.0000

Moving Average
Length: 6

Accuracy Measures
MAPE:       9.3
MAD:      136.0
MSD:  43836.1

 Row   Period  Forecast      Lower      Upper

   1      616      4851   4440.63    5261.37
```

In the Session Window and on the graph, we find the three summary statistics characterizing the historical accuracy of this model. These measures are:

- **MAPE**: Mean Absolute Percentage Error, representing the mean forecast error, expressed as a percentage of the actual observed values. This number is fairly easy to interpret: In this case, on average, our forecast values were "off" by about 9.3% from the actual values.
- **MAD:** Mean Absolute Deviation, simply representing the mean of the absolute values of the forecast errors, in this case $136 million.
- **MSD:** Mean Squared Deviation, which is the sum of the squared forecast errors, divided by *n*. This is very similar to MSE, excepts that MSE divides by the number of forecasts.

In general, we are looking for a Moving Average Length which minimizes these measures of forecast error. Let's generate two more analyses, and select the one with the best error statistics.

🖰 Edit the moving average dialog, and specify a 3-month average period. Change the graph title accordingly.

🖰 Edit the dialog again, this time for a 12-month analysis.

Print all three graphs. ***Which of the three averaging periods is best? What is the forecast for April, 1996 using the best model?***

Forecasting Using Exponential Smoothing

One drawback of moving averages is that, ultimately, our forecast is based on a very few numbers. We discard most of the sample to generate a single forecast. Exponential smoothing manages to incorporate the entire dataset into each forecast. Like moving averages, this technique smoothes out a jagged time series, and does so by creating forecasts which represent a weighted average of the prior forecast and the prior observation. Our job is to select the weights, using a similar strategy as we just did.

Unfortunately, the exponential smoothing command requires a complete set of data, and our series begins with a string of missing values (*'s). So, before we can analyze the series, we must eliminate the missing values. The easiest method is to copy the observations to a new column.

Manip ➤ Copy Columns... In **Copy from columns**, choose `A0M604`, and in **To columns**, type `AgExp`. Then click **Omit Rows**, and specify that we want to omit rows in which the **Numeric Variable** `A0M604` is equal to *.

Stat ➤ Time Series ➤ Single Exp Smoothing... Select the new variable, `AgExp`, and specify a 'smoothing constant' of 0.5. Generate one forecast, and also title the graph appropriately.

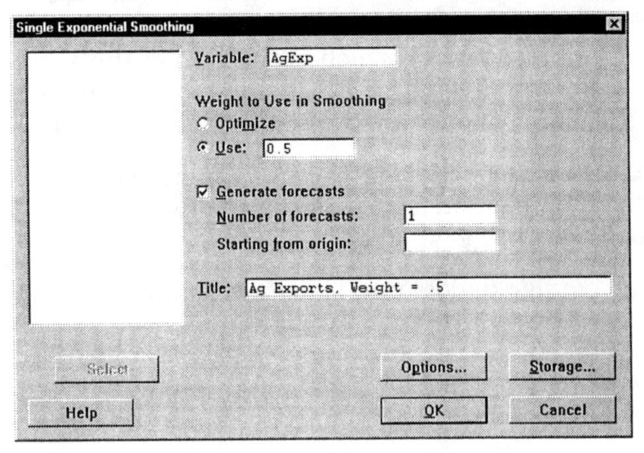

The output is very similar to that of the Moving Average analysis. In terms of MAPE, this choice of a smoothing weight looks relatively good. We could try several more, or we can request that the software optimize the choice of weight for us. Let's do that.

> Edit the last command dialog, and click the option button marked **Optimize**. Change the title of the graph, and click **OK**.

This graph is the best we can do with Exponential Smoothing. ***How do these results compare to the moving average which you chose? How do the April, 1996 forecasts compare to each other?***

Forecasting Using Trend Analysis

It is clear from the graphs that there has been a general upward trend in the dollar value of agricultural exports since the late 1940's. Let's try to model that trend in two ways.

> **Stat ➤ Time Series ➤ Trend Analysis...** Select `A0M604` again, specify a **Linear** trend, and ask for one forecast value.

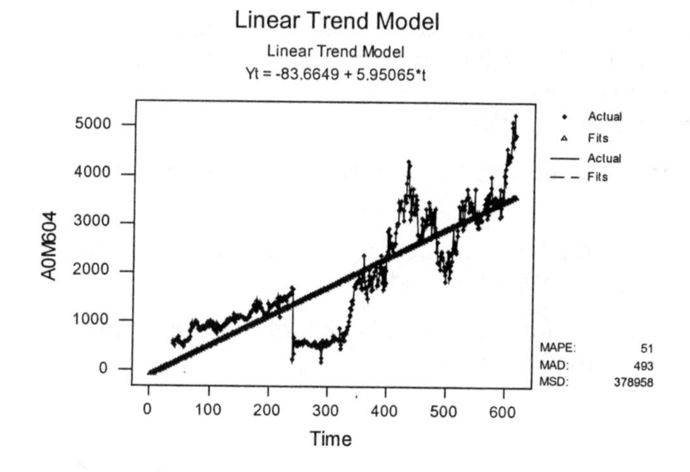

From this graph, we immediately see that the linear trend model is a poor choice. The summary statistics are weak, the actual series oscillates around the trend line, and the forecast for April, 1996 looks very unlikely. Note, however, how clearly you can see the cyclical pattern in the data.

Let's try some non-linear trends.

Edit the trend analysis dialog twice, first choosing the **Quadratic** model, and then the **Exponential Growth** model.

Which of the resulting analyses is better? Why did you select that one?

Moving On...

Using the techniques presented in this lab, answer the following questions.

Output

These questions focus on A0M047, which is an *index* of total industrial production in the United States, much like the Consumer Price Index is a measure of inflation.

1. Which of the four components of a time series can you see in this series? Explain.
2. Generate 6-, 9-, and 12-month moving averages for this series. Which averaging length would you recommend for forecasting purposes, and why? Generate one forecast using that averaging length.
3. Compute one optimal Exponential Smoothing forecast for this variable.
4. Select a trend model which you feel is appropriate for this variable, and generate a forecast.
5. Compare your various forecasts. Which of the predictions would you rely upon, and why?

Utility

6. Generate a 3-month moving average forecast for the variable KWHpDay, which is the mean daily amount of electricity usage in my home. Given the pattern in the entire graph, explain why it may be unwise to rely on this forecast.
7. Why might it be unwise to use a few recent months of data to predict next month's usage? What might be a better approach?

Us

8. Perform a Trend Analysis on the variable `M1`, using Linear, Quadratic, Exponential, and S-Curve models. Which model seems to fit the data best? Why might that be so?

9. Find another variable in this dataset which is well-modeled by the same function as `M1`. Why does this variable have a shape similar to `M1`?

Session 20

Nonparametric Tests

Objectives

In this session, you will learn to do the following:

- Perform a Sign test for a median
- Perform a Wilcoxon Signed Rank test for a median
- Perform a Mann-Whitney U test comparing two medians
- Perform a Kruskal-Wallis test comparing two or more medians
- Compute a Spearman Rank Order correlation coefficient
- Perform a Runs test for randomness

Nonparametric Methods

The previous labs have illustrated various statistical tests involving population parameters (like μ), and which often require the assumption that the underlying population is normally distributed. Sometimes we cannot assume normality, and sometimes the data we have do not lend themselves to computing a mean (for example, when the data are merely ordinal).[1] The techniques featured in this session are known as *nonparametric methods*, and are applicable in just such circumstances. In particular, we will mostly use them when we cannot assume that our sample is drawn from a normal population.

Generally, it is preferable to use a parametric test when the population is close to normal, since such tests tend to be more discriminating and powerful. Using a nonparametric test inappropriately

---

[1] Consult your text to define or review the *scales of measurement* of data.

increases the risk of a Type II error. Despite that risk, nonparametric tests are easy to use and interpret, and free us from the small-sample restrictions on non-normal population data. Although there are a large number of nonparametric methods, this session will illustrate a few of the more common elementary techniques.

A Sign Test

Perhaps the simplest of these techniques is the sign test. We use it to test hypotheses concerning the median of a population. The test relies on the idea that about half of the observations in a random sample should lie above the population median. Thus, with respect to a hypothetical value of the median, the sample is a binomial experiment of *n* trials, each with a 0.5 probability of success, and each independent.

An example will illustrate. You may recall the dataset containing the cholesterol readings for heart attack patients. In addition to the heart attack patients, there was a healthy control group. We'll begin by testing whether the median cholesterol for the control group was less than 200. Following the logic just outlined, if 200 is indeed the population median, we expect about half of the sample observations to be above 200. Formally, we'll test:

H_0: Median $\geq$ 200 vs.
H_A: Median < 200

This is similar to a test we performed in Session 11, but there are two important differences. There, we needed to assume that the population was normally distributed, and that test concerned the mean of the population.

File ➤ Open Worksheet... Open the file **Cholest**.

Stat ➤ Basic Statistics ➤ Display Descriptive Statistics... Select the `Control` variable, and ask for the **Graphical Summary**. This will shed light on the assumption of a normal population.

In the resulting output (see next page), there is some question about whether or not the population is normally distributed, and with a sample of just 30, the t-test might be inappropriate. Therefore, this is a candidate for a sign test.

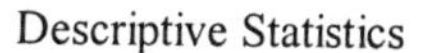

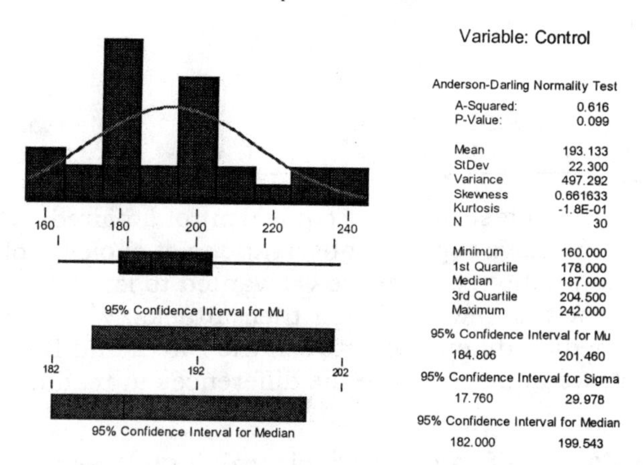

Stat ➤ Nonparametrics ➤ 1-Sample Sign... In the dialog, select `Control`, specify a median of 200, and select a **less than** alternative hypothesis.

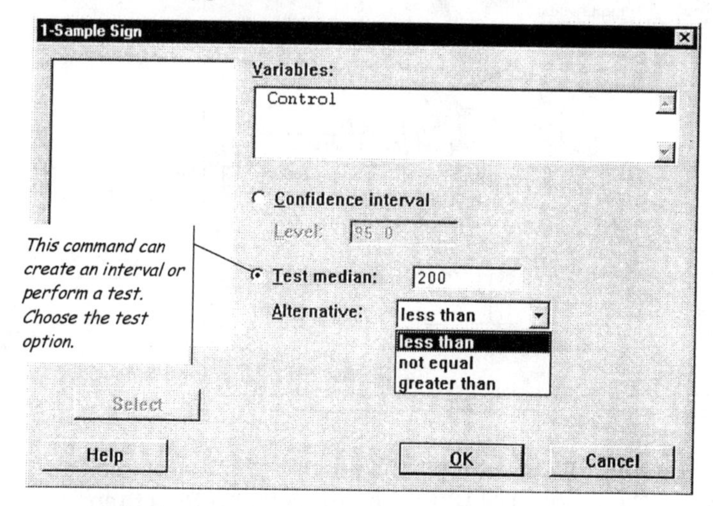

The Session output (see next page) shows that 20 observations were below 200, 2 were equal to 200, and only 8 observations were above it. If the null hypothesis were true, this would happen by chance less than 2% of the time ($p \approx 0.0178$). This all suggests that we are looking at strong evidence that the population median is less than 200.

Sign Test for Median

```
Sign test of median = 200.0 versus  <  200.0

                N  Below  Equal  Above        P    Median
Control        30     20      2      8   0.0178     187.0
```

We can also use the sign test to do the equivalent of a paired-sample t-test. In this case, we have repeated observations of cholesterol readings for the heart attack patients. Suppose we wanted to know if their readings declined in the first few days after their attacks. We might want to test whether the median difference between the Day 2 and Day 4 readings is negative. First, we must compute the differences in readings for each patient.

Calc ➤ Calculator Specify a new variable (call it `Change`), which equals `4-Day - 2-Day` and click **OK**.

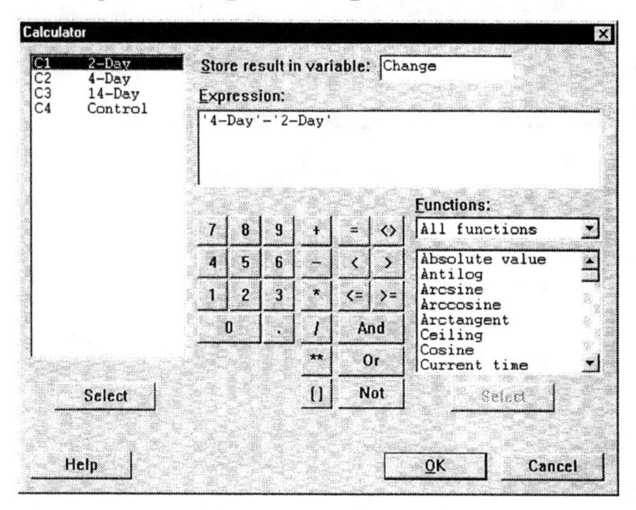

Stat ➤ Nonparametrics ➤ 1-Sample Sign... In the dialog, select `Change`, specify a median of 0, and again select a **less than** alternative. This will test whether the median change was negative (i.e., whether more than half of the patients experienced a decline in cholesterol).

Sign Test for Median

```
Sign test of median = 0.00000 versus  <  0.00000

                 N  Below  Equal  Above         P      Median
Change          28     19      0      9    0.0436      -19.00
```

Here, the results are consistent with the alternative hypothesis: about twice as many patients (19 compared to 9) had decreased cholesterol levels. The p-value of .0436 indicates that we reject the null hypothesis at the 0.05 significance level.

A Wilcoxon Signed Rank Test

In the Sign test, all comparisons are made on the basis of the *sign* of the observation: Is it above or below the median? As far as the Sign Test is concerned, the observation closest to, but above, the median is equivalent to the observation furthest above. The Wilcoxon Signed Rank test (also known as the *Signed Rank Test*) takes the magnitude of the deviations into account by ranking the extent of them. It requires a reasonably symmetric (though not necessarily normal) population.

In this test, the idea is that when we sample from a symmetric population, we should find that the observations are dispersed symmetrically about the population median. This test looks for that symmetry by comparing the rankings of sample observations on each side of the median. It does so by computing all of the individual deviations from the median, ranking them from smallest to largest in absolute value, and dividing them into positive (above the median) and negative. Then it sums the positive and negative ranks, and compares the two sums. If the population is symmetric, and the hypothesized median is correct, the two sums should be very close. In practice, it's simpler than it sounds. Let's apply this test to our first set of data.

Stat ➤ Nonparametrics ➤ 1-sample Wilcoxon... Just as with the Sign test, select the variable `Control`, and test for a **Median** equal to `200` against a **less than** alternative. The results are shown on the next page:

Wilcoxon Signed Rank Test

```
Test of median = 200.0 versus median  <  200.0

                N for   Wilcoxon             Estimated
           N    Test  Statistic          P      Median
Control   30      28      127.5      0.044       191.0
```

In this test, two of the thirty observations are omitted because they are equal to the hypothesized median value of 200. Based on the remaining 28, the procedure computes the test statistic of 127.5, which is significant at the 0.05 level (p = .044).[2] Thus, taking ranks into account, we reject the null once again, but with less confidence.

We can also use the Wilcoxon Ranked Sign test as an equivalent to the paired t-test, just as we did the sign test. As an exercise, perform a Wilcoxon Ranked Sign to test the null hypothesis that cholesterol levels remain stable or increase from Day 2 to Day 4 among heart attack patients. ***Compare the results of this Wilcoxon test to those of the sign test. Comment on noteworthy similarities and differences between the results of the two tests.***

Mann-Whitney U Test

This test is analogous to the t-test for two independent samples, but once again is a test of medians. In particular, we use this test when we have two independent samples and can assume that they are drawn from populations with the same (but not necessarily normal) shape and variance.

Like the previous tests, the U Test can be used for ordinal, interval, or ratio data, and is based on rankings. The underlying strategy is similar to the Wilcoxon Ranked Sum test. If the medians of the two populations are equal, and the shape of the two populations is the same, then when we pool and rank all of the observations, the rankings should be balanced across the two samples.

Recall the data from the Salem Witchcraft Trials. In that dataset, we have information about the taxes paid by each individual in Salem Village. We can also distinguish between those who were supporters of the Village minister, Reverend Parris, and those who were not. In this

---

[2] Consult your text or Minitab help for the computational details.

test, we will hypothesize that supporters and non-supporters paid comparable taxes—or that the median tax was equivalent in both groups.

Before performing the test, we want to check the assumptions of same shape and variance. We can do so by consulting the descriptive statistics. Open the worksheet called **Salem**.

> **Stat ➤ Basic Statistics ➤ Display Descriptive Statistics...** Select the variable `Tax`, and check **By variable**. The grouping variable is `ProParris`. Also, click on **Graphs**, and request the **Graphical Summary**.

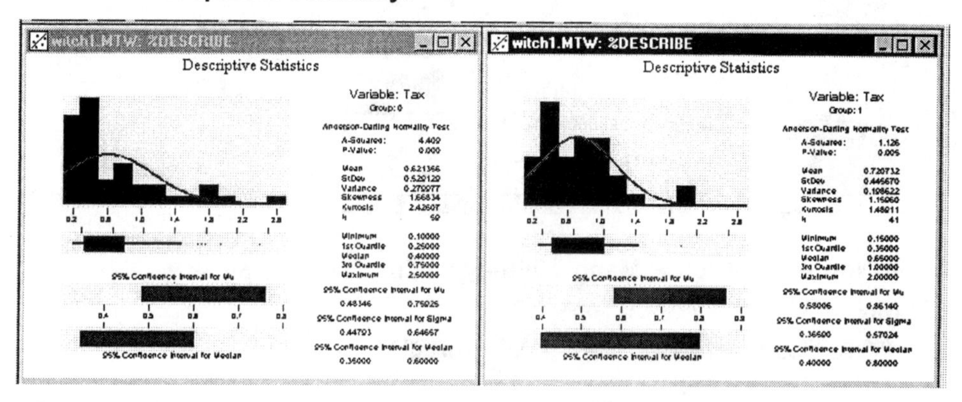

Here we see two skewed distributions of comparable shape, which are clearly *not* normal. Their variances are similar, and so meet the requirements of the U test more nearly than those of the two-sample t-test.

In order to perform a Mann-Whitney test in Minitab, our samples must be in separate columns of the worksheet; that is, they must be "unstacked."[3] In our current worksheet, an individual's tax payment and political stance concerning the minister are represented as two variables. Because of the way the Mann-Whitney command works, we need to think of the political stance as differentiating two separate populations, and create two new columns of tax payment data, representing the pro- and anti-Parris factions in the town. We do so as follows:

> **Manip ➤ Stack/Unstack ➤ Unstack One Column...** As shown in the dialog box on the next page, specify that you want to **Unstack the data in** `Tax`, separating it into two new columns

[3] For more information about "stacked" and "unstacked" data, see Appendix C.

called **TaxAnti** and **TaxPro**, using the 0–1 subscript data in **ProParris**.

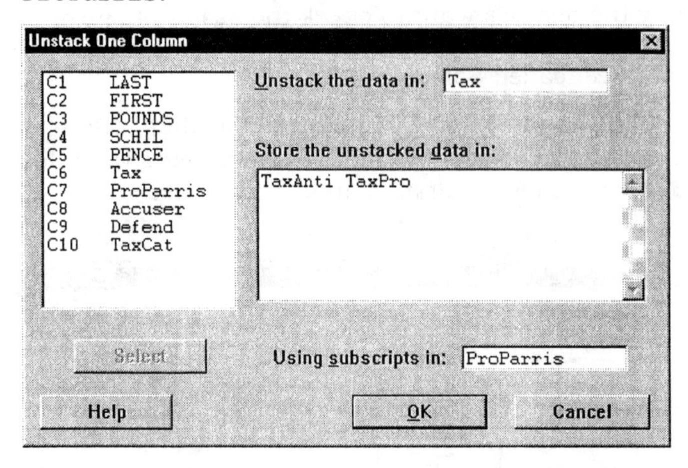

Stat ➤ Nonparametrics ➤ Mann-Whitney... The first sample is **TaxAnti**, and the second is **TaxPro**.

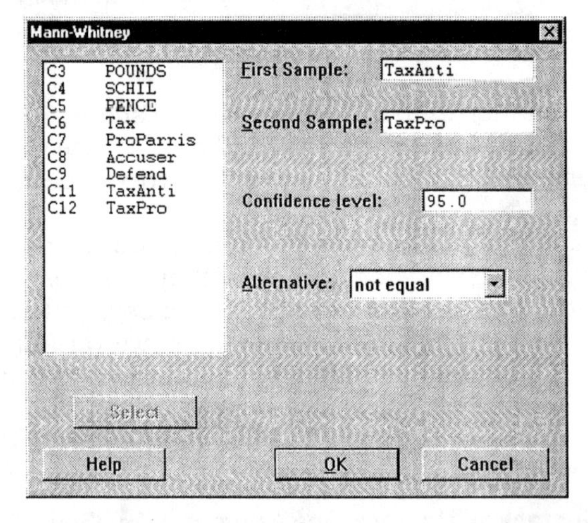

The results (Session Window; see next page) display the sample medians for the two groups, and the difference between them (-0.15). ETA1 and ETA2 refer to the respective population medians. The test statistic (W) is the sum of the rankings from the first sample.

```
Mann-Whitney Confidence Interval and Test

TaxAnti     N =  59     Median =       0.4000
TaxPro      N =  41     Median =       0.6500
Point estimate for ETA1-ETA2 is       -0.1500
95.0 Percent CI for ETA1-ETA2 is (-0.3001,0.0001)
W = 2694.5
Test of ETA1 = ETA2  vs  ETA1 not = ETA2 is significant at 0.0462
The test is significant at 0.0454 (adjusted for ties)
```

Minitab computes two p-values and reports whether or not the test is significant at the 0.05 level. The first p-value is based on all of the sample data; the second adjusts the p-value in the event that there were ties in the rankings between the two samples. The unadjusted p-value is conservative (i.e., higher) when there are ties; the adjusted p-value is, however, the more accurate of the two when there are ties. In this example, the test is significant, which is to say that the sample data indicate a statistically significant difference between the median taxes paid by pro- and anti-Parris residents.

Kruskal-Wallis Test

Whereas the Mann-Whitney test addresses the comparison of central location for two non-normal populations, the next test does the same for *two or more* non-normal populations.

As in the Mann-Whitney test, the Kruskal Wallis H-test assumes that the samples are drawn from independent populations with identical shape and spread. Unlike the one-way ANOVA, there is no need to assume normality. The strategy is once again based on a comparison of pooled rankings, and the null hypothesis is that the medians of the k populations are equal; the alternative hypothesis is that at least one of the medians is different.

To illustrate, open the beverage industry worksheet (**Bev**). This dataset contains financial data about 82 firms in 5 different segments of the beverage industry. We will focus on the variable called `Current`. A firm's *current ratio* is the ratio of its current assets to current liabilities. A firm with a current ratio of 1 could quickly cover all of its current liabilities by converting assets to cash. In this illustration we'll ask whether firms in different parts of the beverage industry have the same median current ratio. The firm's industry segment is identified by Standard Industry Classification Code (see Appendix A for definitions).

Stat ➤ Nonparametrics ➤ Kruskal-Wallis... The terminology here is reminiscent of ANOVA. The **Response** variable is `Current`, and the **Factor** is `SIC` code.

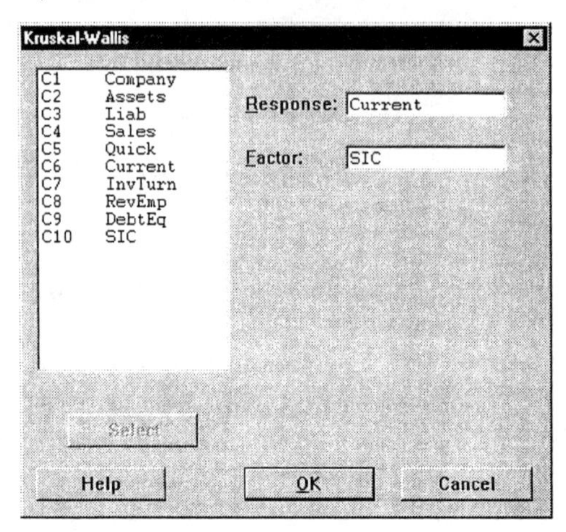

Once again, the results are in the Session Window:

Kruskal-Wallis Test

```
Kruskal-Wallis Test on Current

SIC          N     Median     Ave Rank          Z
2082        27      1.170         49.1       0.72
2084        10      2.185         64.0       2.28
2085         6      1.860         52.1       0.58
2086        35      1.050         37.2      -2.51
2087        13      1.710         46.7       0.10
Overall     91                    46.0

H = 9.22  DF = 4  P = 0.056
H = 9.22  DF = 4  P = 0.056 (adjusted for ties)
```

In the output, we see the median Current ratios for each SIC code, as well as the mean ranking for each. The test statistic, which follows a Chi-square distribution with *k-1* degrees of freedom, in this case is H = 9.22, which has a p-value equal to 0.056. As in the Mann-Whitney test, the test statistic is also adjusted in the event of tied rankings (no ties in this example). The conclusion in this particular example depends on our choice of significance level.

Spearman Rank Order Correlation

In one of the earliest sessions, we learned to compute the Pearson correlation coefficient, which measures the linear association between two interval or ratio variables. Sometimes, we have only ordinal data but may still suspect a linear relationship. For example, we may want to compare the order of finish for runners in two qualifying races.

The Spearman Rank Order Correlation coefficient is simply the familiar Pearson *r*, applied to ordinal data. For example, recall the data about the 50 states (**State**). In that dataset, we have the average wages for each state in 1993 and 1994. Suppose we wanted to determine the extent to which the rankings of the states changed, if at all, in one year. We'll first rank the states for each of the years, and then compute the correlation coefficient for the ranks.

- **Manip ➤ Rank** As shown in the dialog, we want to rank the states by `Pay93`, and place the rankings into a new column which we'll label `PayRank93`. Each state will be assigned a rank, with 1 assigned to the lowest-wage state (South Dakota).

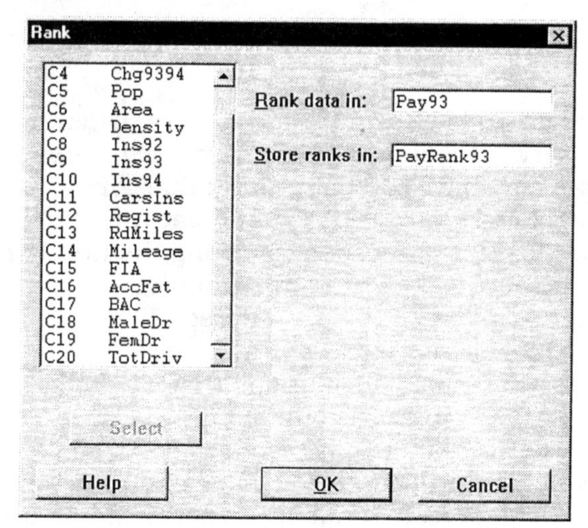

- Now rank the states by `Pay94`, creating a new column called `PayRank94`.

A brief look at the two new columns in the Data Window suggests that there were slight changes in the rankings. Let's see how close the two lists are.

🖱 **Stat ➤ Basic Statistics ➤ Correlation...** This is a familiar dialog. Select the two pay rank variables, and click **OK**.

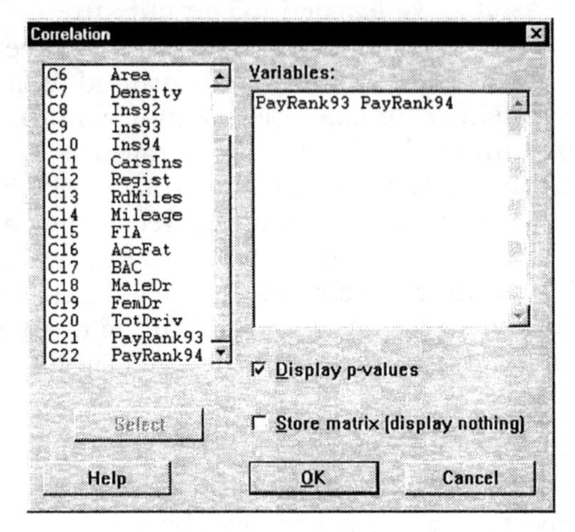

In the Session Window, you'll find that the correlation is 0.998, which is significantly different from zero.

A Runs Test

The final test in this session is a test for randomness. Often, we may want to ask whether observations in a sequence are independent of one another. For instance, we may have reason to think that an observation at time t was correlated with the observation at time $t - 1$. The Runs test checks a series for randomness by counting the number of "runs" in the data. A run is defined as a consecutive group of observations sharing a particular characteristic of interest, like consecutive heads or tails when flipping coins.

We know from studying binomial distributions that we can reasonably expect occasional runs in any Bernoulli process with independent trials. However, we also know that there are probability distributions pertaining to such patterns, and that some run patterns would stretch credulity that the data were randomly generated. The Runs test is based on that fact.

Let's consider two examples. The first will use simulated data, and we'll use the test to see if indeed Minitab's random data generator creates a random sequence.

🖰 **Calc ➤ Random Data ➤ Bernoulli...** Generate 100 rows of random data, with a probability of success = .5. This will create a randomly-generated column of 0's and 1's. Store the results in C25.

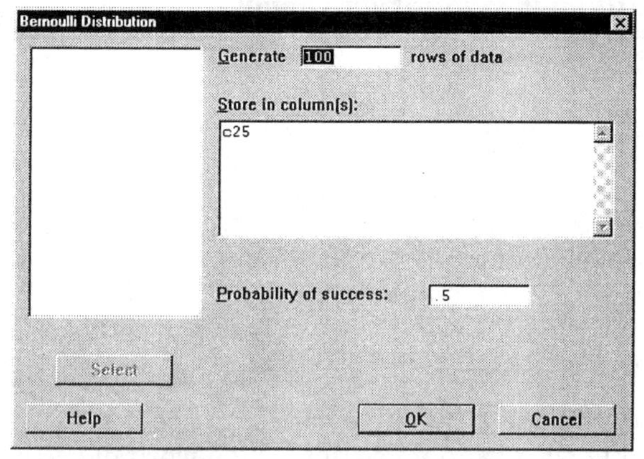

🖰 In the Data Window, label C25 as **`Bernoulli`**.

🖰 **Stat ➤ Nonparametrics ➤ Runs Test...** Select **`Bernoulli`**.

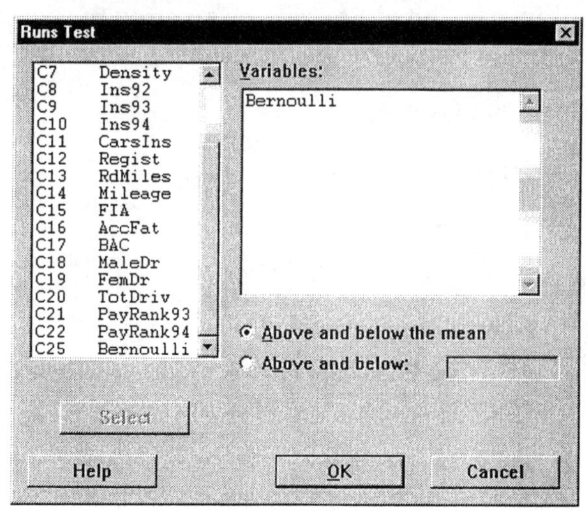

The Runs test output appears in the Session Window; naturally, your random data and test results will differ from those shown in the illustration on the next page. The test compares the expected number of runs in a series of 100 observations to the observed number, and computes a significance level. In this example, the value K = .51 is the sample mean, and 51 observations lay above it. We expected 50.98 runs and observed 45. The null hypothesis of a random arrangement cannot be rejected, which is to say that we persist in our assumption of a random sequence. ***What did your simulation show?***

Runs Test

```
    Bernoull
    K =      0.5100

    The observed number of runs =  45
    The expected number of runs =  50.9800
    51 Observations above K    49 below
              The test is significant at  0.2291
              Cannot reject at alpha = 0.05
```

The second example will use a real time series, to see if a particularly "noisy" series follows a random pattern. One word of warning when applying the Runs Test in Minitab: Your variable may have missing cases at the beginning or end of the series, but *not* anywhere in the middle. In this example, the variable is the percentage of industrial capacity utilized each month in the United States over a period of years. To take a look at the series, first open the worksheet called **Output**. The variable of interest is called `A0M124`.

The time series itself looks like this:

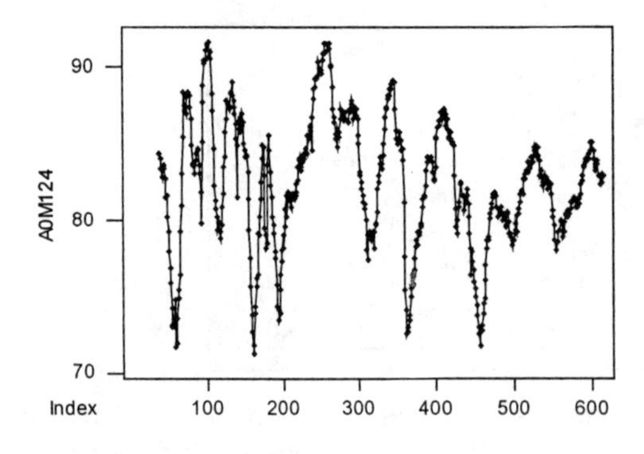

The series is quite erratic, and we might hypothesize that it is a random series. The Runs test can evaluate that hypothesis.

Stat ➤ Nonparametrics ➤ Runs Test... Select `A0M124`.

Based on this test, how many of the observations were above and below the sample mean? Does this test conclude that the time series is or is not random?

Moving On...

Apply the techniques learned in this session to the questions below. In each case, indicate which test you have used, and cite appropriate test statistics and p-values.

Output

1. Is the capacity utilization among manufacturing firms (`A0M082`) a random series?
2. Use two different approaches to test the hypothesis that the median manufacturing capacity utilization is 80%. Would a t-test be more or less appropriate than a nonparametric test in this case? Explain.

Aids

This file contains data about the number of AIDS cases from 208 countries around the world.

3. Was there any significant change in the median *rate* of AIDS cases from 1992 to 1993?
4. Compute the rankings of the countries in terms of number of cases reported in 1992 and 1994. Compute a Spearman correlation coefficient, and interpret the coefficient.
5. Did all WHO regions experience roughly the same rate of cases in 1993?

F500

These are the Fortune 500 companies for 1996.

6. The first two variables rank the companies by total revenue for 1994 and 1995. Compute and interpret the rank

correlation coefficient. What does this suggest about referring to a particular company as “a Fortune 500 firm?”

7. The variable called ROI8595a is the annual rate of return to investors from 1985–1995. Should investors expect a median return in excess of 12%, assuming that these years are representative of future returns?

Swimmer1

8. Did swimmers tend to improve between their first and last recorded time, regardless of event?

9. Looking only at second race data for the 50-meter events, are the median times the same regardless of event? [Hint: You'll need to use Subset or Copy to isolate the data you want to analyze.]

Session 21

Statistical Process Control

Objectives

In this session, you will learn to do the following:

- Create and interpret a mean control chart
- Create and interpret a range control chart
- Create and interpret a standard deviation control chart
- Create and interpret a proportion control chart

Processes and Variation

We can think of any organizational or natural system as being engaged in *processes*, or series of steps and interactions which produce an outcome. In organizations, goods and services are the products of processes.

One dimension of product or service quality is the minimization of process variation. That is to say, one difference between goods of higher or lesser quality often is their *consistency*. Those who are responsible for overseeing a process need tools for detecting and responding to variation in a process.

Of course, some variation may be irreducible, or at times even desirable. If, however, variation arises from the deterioration of a system, or from important changes in the operating environment of a system, then some intervention or action may be appropriate.

It is critical that managers intervene when variation represents a problem, but that they avoid unnecessary interventions which either do harm or do no good. Fortunately, there are methods that can help a manager discriminate between such situations.

This lab session introduces a group of statistical tools known as *Control Charts.* A control chart is a time-series plot of a sample statistic. Think of a control chart as a series of hypothesis tests, testing the null hypothesis that a process in "under control."

How do we define "under control?" We will distinguish between two sources or underlying causes of variation:

- ***Common cause*** (also called *random* or *chance*)—Typically due to the interplay of factors within or impinging upon a process. Over a period of time, they tend to "cancel" each other out, but may lead to noticeable variation between successive samples. Common cause variation is always present.
- ***Assignable cause*** (also called *special* or *systematic*)—Due to a particular influence or event, often one which arises "outside" of the process.

A process is "under control" or "in statistical control" when all of the variation is of the common cause variety. Managers should intervene in a process with assignable cause variation. Control charts are useful in helping us to detect assignable cause variation.

Charting a Process Mean

In many processes, we are dealing with a measurable quantitative outcome. Our first gauge of process stability will be the sample mean, $\bar{x}$. Consider what happens when we draw a sample from a process that is under control, subject only to common cause variation. For each sample observation, we can imagine that our measurement is equal to the true (but unknown) process mean, μ, plus or minus a small amount due to common causes. In a sample of *n* observations, we'll find a sample mean, $\bar{x}$. The next sample will have a slightly different mean, but assuming that the process is under control, the sample means should fluctuate near μ. In fact, we should find that virtually all samples fluctuate within three standard errors of the true population mean.

An $\bar{x}$-chart is an ongoing record of sample means, showing the historical (or presumed) process mean value, as well as two lines representing "control limits." The control limits indicate the region approximately within three standard errors of the mean. Minitab computes the control limits, described further below, and can perform a series of tests to help identify assignable causes. These tests are reliable when our data are approximately normal. An example will illustrate.

Recall the household utility data (in worksheet **Utility**). In this file we have 81 monthly readings of electricity and natural gas consumption in my home, as well as monthly temperature and climate data. We'll start by creating a control chart for the monthly electricity consumption, which varies as the result of seasonal changes and family activity.

We added a room and made some changes to the house, beginning roughly five years along in the dataset. We suspect that the construction project and the presence of additional living space may have increased average monthly electricity usage.

Stat ➤ Control Charts ➤ Xbar... In this worksheet, we have one electricity usage measurement per month, stored in `C8 KWHpDay`. We will "sample" three months at a time, approximating the four seasons within a year. In the dialog box, select the variable `KWHpDay`, and indicate that the 'subgroup size' is 3 months (as shown below). Before clicking **OK**, click on **Tests....**

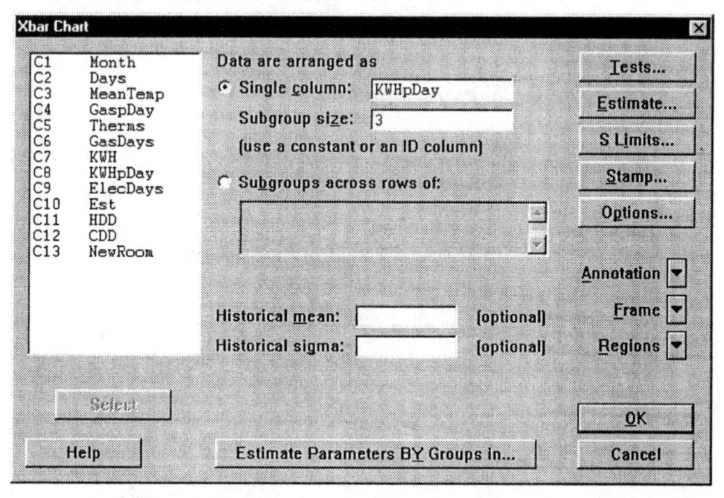

As noted earlier, Minitab will perform a series of tests designed to identify evidence of assignable (or special) causes. The eight are listed in the dialog box illustrated below. To clarify the illustration, I have checked all eight tests, but you should simply select **Perform all eight tests**, and click **OK**, and create the chart.

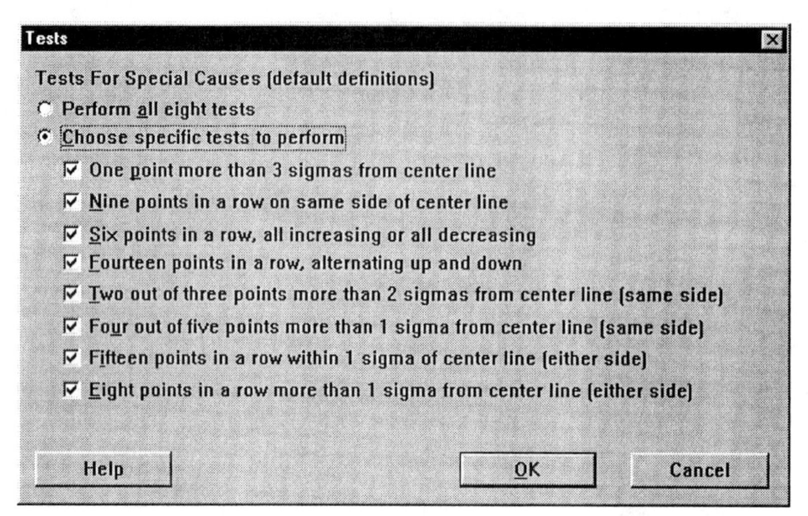

The dialog box describes the tests quite fully. Each test represents a reason to suspect that the sample means may not be random. In the resulting control chart, the points numbered in red indicate places at which a specific test failed. Let's take a close look at this control chart.

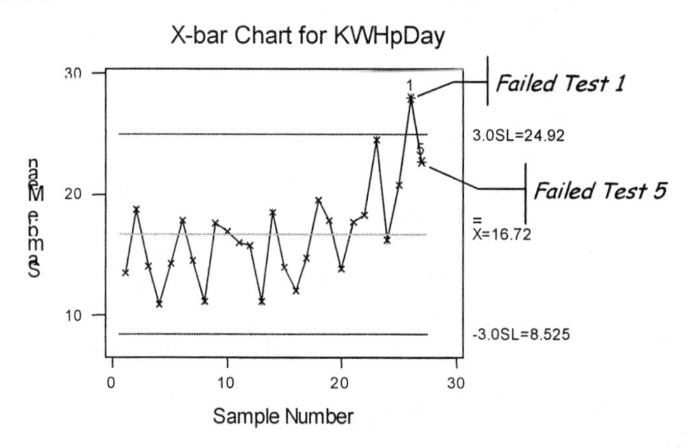

The basic "geography" of the chart is simple. On the horizontal axis is the sample number; we have 27 samples of 3 observations each, collected over time. The vertical axis represents the means of each sample. The green horizontal line, labeled $\bar{\bar{x}}$, is the grand mean of all samples. The two red lines are the Upper and Lower Control (sigma) Limits. The upper line is 3 standard errors above the mean, and the lower is 3 standard errors below. By default, Minitab estimates sigma using the pooled standard deviation from the sample data; other methods are optionally available, but beyond the scope of this session. You should be aware that if there are any missing observations in the data, the control limits will vary from sample to sample, and the tests for special causes will *not* be performed.

What does this chart tell us about electricity consumption? For the first 20 samples, the process appears to be in control. There are no test failures, and the usage appears to oscillate seasonally. Beginning with sample 21 (month 63 and after) we see a disturbance in this pattern. The red X's and numbers on the graph correspond to messages in the Session Window, shown here:

```
TEST 1. One point more than 3.00 sigmas from center line.
Test Failed at points: 26

TEST 5. 2 out of 3 points more than 2 sigmas from center line
        (on one side of CL).
Test Failed at points: 27

TEST 6. 4 out of 5 points more than 1 sigma from center line
        (on one side of CL).
Test Failed at points: 27
```

CL stands for Center Line

Collectively, these tests indicate that between samples 22 and 27, there were an unusually high number of sample means far above the grand mean. This suggests an assignable cause—a construction project requiring power tools, and new living space thereafter.

The Xbar chart and tests presume a constant process standard deviation. Before drawing conclusions based on this chart, we should examine the standard deviations among the samples. What is more, sample variation is another aspect of process stability. We might have a process with a rock-solid mean, but whose standard deviation grows over time. Such a process would also require some attention. Later we'll illustrate the use of a Standard Deviation (S) chart; the S chart is appropriate for samples with 5 or more observations. For smaller samples, such as we have here, the better tool is the Range (R) chart.

Charting a Process Range

The Range chart tracks the sample ranges (maximum minus minimum) for each sample. It displays a mean range for the entire dataset, and control limits computed based upon the mean sample range. The range chart has four tests available for assignable causes.

> **Stat ➤ Control Charts ➤ R...** Just as with the Xbar chart, our data are in a single column. The variable is still **`KWHpDay`**, and the subgroup size is still 3.

> Once again, select **Tests...** and select **Perform all four tests.**

When you have properly completed the dialogs, you should see this graph:

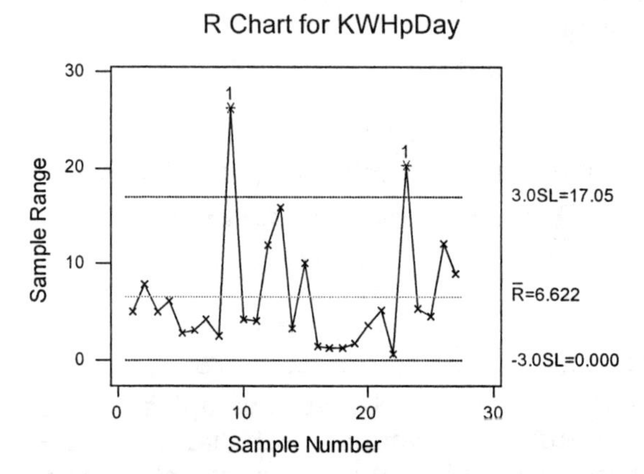

This chart is comparable in structure to the Xbar chart. The major difference is that the lower control limit is closer to the center line than the upper limit. This is because the sample range can never be less then 0; if the computed lower SL were to be negative, as is the case here, it is set at 0. In a stable process, the sample ranges should be randomly distributed within the control limits.

In two samples (9 and 23) the sample range is more than approximately 3 standard errors above the average sample range. This suggests some instability in the process. Something unusual in those two sample periods led to an extraordinary difference between the high and low readings.

It is also important to compare the Xbar and R charts; when a process is under control, both charts should fluctuate randomly within control limits, and should not display any obvious connections (e.g. high means corresponding to high ranges). In practice, we often place the mean and range charts on a single page, as follows.

🖱 **Stat ➤ Control Charts ➤ Xbar-R...** Proceed as you did in the prior dialogs, generating this chart shown on the next page.

In this chart, note that the Xbar chart is slightly different than before. Here, the standard error is estimated based upon the sample ranges rather than the sample standard deviations. This locates the control limits differently, and therefore sets new standards for the tests. In this output, we have further evidence of a process out of control at several points, suggesting that the home owner might want to intercede to stabilize electricity use.

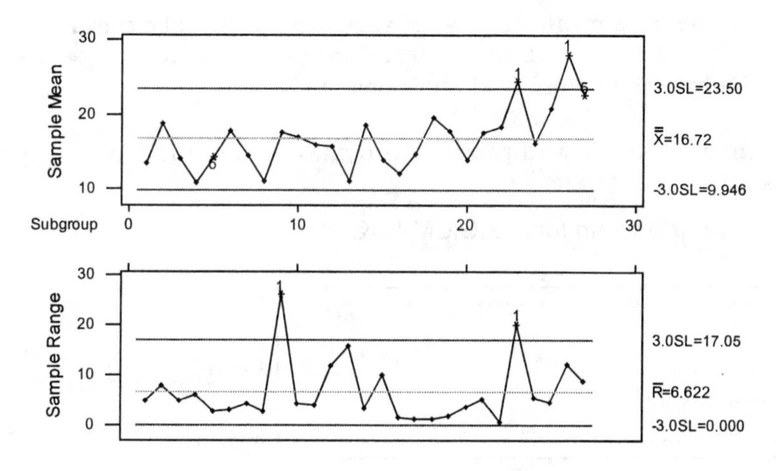

Another Example

In the previous example, the observations were in a single column. Sometimes sample data is organized so that the observations within a single sample are in a row across the worksheet, and each row represents the set of sample observations. To construct control charts from such data, we need only make one change in the relevant dialogs.

To illustrate, open the worksheet called **EuropeY**. This file contains data extracted from the Penn World Tables dataset (**PaWorld**), isolating just the 14 European countries in that dataset. Each row represents one year (1960–1990), and each column a county. Each value in the worksheet is the ratio of a country's annual real per capita Gross Domestic Product to the per capita GDP of the United States, expressed as a percentage. A value of 100 means that real per capita GDP for the country was identical to that in the U.S. A variety of economic and political factors presumably influence a nation's income relative to the United States; thus we may conceive of these figures as arising from an ongoing process.

Stat ➤ Control Charts ➤ Xbar-S... Since we have 14 observations per sample, let's use the standard deviation rather than the sample range as the basis of sample variation. Select the button marked **Subgroups across rows of.** Position the cursor in the white box below that label, and click once.

Then move the cursor into the list of variable names; click and drag to highlight the list of names from `Belgium` to `U.K.` Release the mouse button, and click **Select**.

Request all eight tests, and produce the chart. It should look like this:

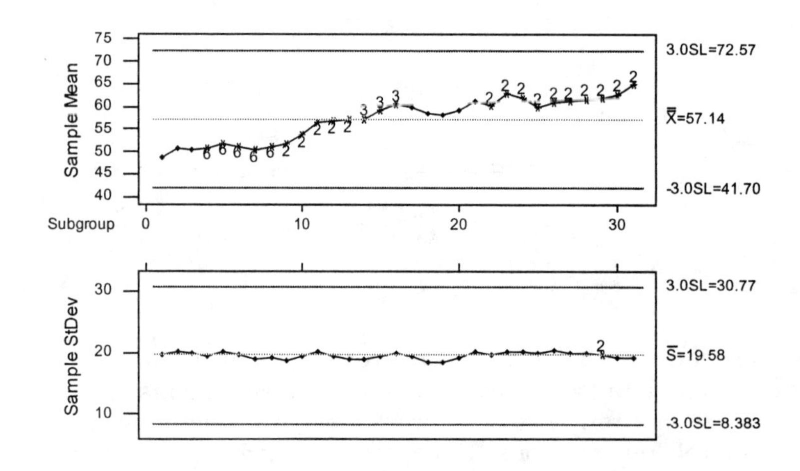

Here we see many test failures in the means, but fairly stable standard deviations. The mean real per capita GDP in Europe has been steadily rising relative to the United States. Statistically, this process is not in control; from the perspective of the European countries, that may be just fine.

Charting a Process Proportion

The previous examples dealt with measurable process characteristics. Sometimes, we may be interested in tracking a qualitative attribute or event in a process, and therefore focus our attention on the frequency of that attribute. In such cases, we need a control chart for the relative frequency or sample proportion. For our example, let's consider the process of doing research on the World Wide Web.

There are a number of "search engines" available to facilitate Internet research. The user enters a keyword or phrase, and the search engine produces a list of Universal Resource Locators (URL's), or Web addresses relevant to the search phrase. Sometimes, a URL in the search engine database no longer points to a valid Web site. In a rapidly changing environment like the Internet, it is not uncommon for Web sites to be temporarily unavailable move or vanish. This can be frustrating whether one is conducting research, job-hunting, or even shopping on the Web.

One very popular search engine is Yahoo!®. Yahoo! offers a feature called the Random Yahoo! Link. When you select this link, one URL is randomly selected from a massive database, and you are connected with that URL. As an experiment, I sampled twenty random links, and recorded the number of times that the link pointed to a site which either did not respond, was no longer present, or had moved. I repeated the sampling process twenty times through the course of a single day, and entered the results into the worksheet called **Web**. Open it now.

> **Stat ➤ Control Charts ➤ P...** This dialog is similar to those before. Select the **Variable** called `Problems`, and indicate that the **Subgroup size** is 20. Click **OK**.

As evidenced in the control chart shown on the next page, this process is under control. All of the variation is within the control limits, and there are no reports of any failed tests. Note that the lower control limit is set at 0, since we can't have a negative proportion of problem URLs. The chart indicates that approximately 12% of the attempted connections encountered a problem of some kind, and that proportion remained stable through the course of a single day.

In this example, all samples were the same size (n=20). Had they been of different sizes, we would need a variable indicating the sample size, and would identify that variable in the **Subgroups in** portion of the dialog. When subgroups vary in size, the control limits are different for each sample, and no tests are performed.

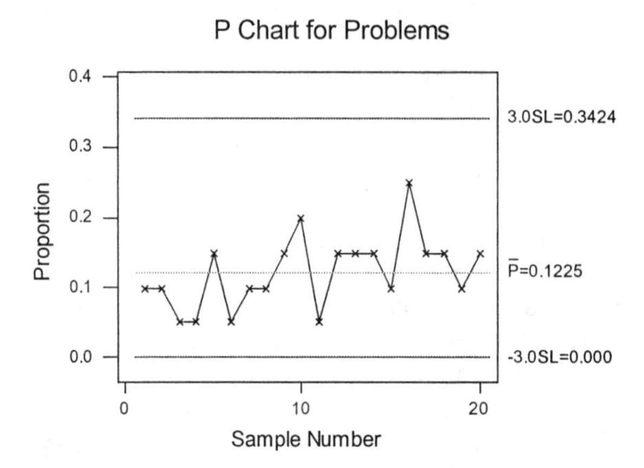

Moving On...

Use the techniques presented in this session to examine the processes described below. Construct appropriate control charts, and indicate whether the process appears to be in control or not. If not, speculate about the possible assignable causes which might account for the patterns you see.

Web

1. You can repeat my experiment with the Random Yahoo! Link, if you have access to a Web browser. In your browser, establish a bookmark to this URL:

   http://random.yahoo.com/bin/ryl

   Then, each time you select that bookmark, a random URL will be selected, and your browser will attempt to connect you. Tally the number of problems you encounter per 20 attempts. Repeat the sampling process until you have sufficient data to construct a p-chart. NOTE: This process can be very time-consuming, so plan ahead.

   Comment on how your p-chart compares to mine.

London1

The National Environmental Technical Center in Great Britain continuously monitors air quality at many locations. Using an automated system, the Center gathers hourly readings of various gases and particulates, twenty-four hours every day. The worksheet called **London1** contains a sampling of the hourly measurements of carbon monoxide (CO) for the year 1996, recorded at a West London sensor.[1] The historical (1995) mean CO level was .669, and the historical sigma was .598.

2. Chart the sample means for these data. Would you say that this natural process is under control?
3. Do the sample ranges and sample standard deviations appear to be under control?
4. The filed called **London2** contains all of the hourly observations for the year (24 observations per day, as opposed to 6). Repeat the analysis with these figures, and comment on the similarities and differences between the control charts.

Labor

5. The variable called `A0M091` is the mean duration of unemployment as of the observation month. Using a subgroup size of 12 months, develop appropriate control charts to see whether the factors affecting unemployment duration were largely of the common cause variety.
6. The variable called `A0M001` represents the mean weekly hours worked by manufacturing workers in the observation month. Using a subgroup size of 12 months, develop appropriate control charts to see whether the factors affecting weekly hours were largely of the common cause variety.
7. Using the Calculator, create a new variable called **`EmpRate`**, equal to total Civilian Employment divided by Labor Force. This will represent the *proportion* of the labor force which was actually employed during the observation month. Using a subgroup size of 12 months, develop appropriate control

---

[1] The file contains 325 days of data, with 6 observations at regular 4-hour intervals each days. Apparently due to equipment problems, some readings were not recorded.

charts to see whether the factors affecting employment were largely of the common cause variety.

Eximport

8. Using the Calculator, create a new variable called `Ratio`, equal to total Exports divided by Total Imports. Using a subgroup size of 12 months, develop appropriate control charts to see whether the factors affecting the ratio of exports to imports were largely of the common cause variety.

EuropeC

9. The data in this worksheet represent consumption as a percentage of a country's Gross Domestic Product. Each row is one year's data from 14 European countries. Using the rows as subgroups, develop appropriate control charts to see whether the factors affecting consumption were largely of the common cause variety.

Appendix A

Dataset Descriptions

This appendix contains detailed descriptions of all Minitab worksheets provided with this book. Refer to it whenever you work with the datasets.

AIDS.MTW

Reported cases of Acquired Immune Deficiency Syndrome (AIDS) from 208 countries around the world. N = 208. Source: World Health Organization, 1995.
Gopher://itsa.ucsf.edu:70/11/.i/.q/.d/library/worldstats/950118 worldstats8.

Col.	Var.	Description
C1	Country	Name of Country
C2	WHOReg	World Health Org. Regional Office AFRO = Africa AMRO = Americas EMRO = Eastern Mediterranean EURO = Europe SEARO = South-East Asia WPRO = Western Pacific
C3	Case7991	Number of cases reported, 1979–1991
C4	Rate92	Cases per 100,000 population, 1992
C5	Case92	Number of cases reported, 1992
C6	Case93	Number of cases reported, 1993
C7	Rate93	Cases per 100,000 population, 1993
C8	Case94	Number of cases reported, 1994
C9	TotCase	Total Number of cases reported, 1979–1994

AIDS2.MTW

Total number of AIDS cases reported, 1979–1994, by continent. N = 5. Source: World Health Organization, 1995. Gopher://itsa.ucsf.edu:70

Col.	Var.	Description
C1	Region	Name of Continent
C2	Country	Number of Countries reporting
C3	TotCases	Cumulative Number of cases, 1979–1994

ANSCOMBE.MTW

This is a dataset contrived to illustrate some hazards of using simple linear regression in the absence of a model or without consulting scatterplots. N = 10. Source: Anscombe, F.J., "Graphs in Statistical Analysis," *The American Statistician*, v. 27, no. 1 (February, 1973), pp. 17–21. Used with permission.

Col.	Var.	Description
C1	X1	First independent variable
C2	Y1	First dependent variable
C3	X2	Second independent variable
C4	Y2	Second dependent variable
C5	X3	Third independent variable
C6	Y3	Third dependent variable
C7	X4	Fourth independent variable
C8	Y4	Fourth dependent variable

BEV.MTW

Financial data from 91 firms in the beverage industry in fiscal 1994. N = 91. Sources: Company reports and data extracted from Compact Disclosure © 1996, Disclosure Inc.

Col.	Var.	Description
C1	Company	Name of Company
C2	Assets	Total Assets (000's of $)
C3	Liab	Total Liabilities (000's)
C4	Sales	Gross Sales
C5	Quick	Quick Ratio ({current assets–inventory}/current liab)
C6	Current	Current Ratio (current assets/current liab)
C7	InvTurn	Inventory Turnover
C8	RevEmp	Revenue per Employee
C9	DebtEq	Debt to Equity Ratio

C10	SIC	Principal Standard Industry Classification Code (SIC) 2082 Malt Beverages 2084 Wines, Brandy, & Brandy Spirits 2085 Distilled & Blended Liquors 2086 Bottled & Canned Soft Drinks 2087 Flavoring, Extract & Syrup

BODYFAT.MTW

Various body measurements of 252 males, including two estimates of the percentage of body fat for each man. Sources: Penrose, K., Nelson. A, and Fisher A., "Generalized Body Composition Prediction Equation for Men Using Simple Measurement Techniques," *Medicine and Science in Sports and Exercise* , v. 17, no. 2, p. 189 (1985). Used with permission. Available via the *Journal of Statistics Education.* http://www.stat.ncsu.edu/info/jse, contributed by Prof. Roger Johnson of South Dakota School of Mines and Technology.

Col.	Var.	Description
C1	Density	Body density, measured by underwater weighing
C2	FatPerc	Percent body fat, estimated by Siri's equation[1]
C3	Age	Age, in years
C4	Weight	Weight, in pounds
C5	Height	Height, in inches
C6	Neck	Neck circumference (cm)
C7	Chest	Chest circumference (cm)
C8	Abdomen	Abdomen circumference (cm)
C9	Hip	Hip circumference (cm)
C10	Thigh	Thigh circumference (cm)
C11	Knee	Knee circumference (cm)
C12	Ankle	Ankle circumference (cm)
C13	Biceps	Biceps circumference (extended; cm)
C14	Forearm	Forearm circumference (cm)
C15	Wrist	Wrist circumference (cm)

CENSUS90.MTW

This is a random sample of raw data provided by the Data Extraction System of the U.S. Bureau of the Census. The sample

[1] Siri's equation provides a standard method for estimating the percentage of fat in a person's body. Details may be found in the libstat web site noted above, or in Siri, W.E., (1956), "Gross composition of the body", in *Advances in Biological and Medical Physics,* v. IV, edited by J.H. Lawrence and C.A. Tobias, Academic Press, Inc., New York.

contains selected responses of 982 Massachusetts residents, drawn from their completed 1990 Decennial Census forms. N = 982. Source: U.S. Dept. of Census, http://www.census.gov

Col.	Var.	Description
C1	Code	{unused}
C2	Age	Age of respondent (years, except as noted below) 0 = Less than 1 year 90 = 90 yrs or more
C3	Avail	Available for work 0 = n/a under 16 yrs old/at work/not looking 1 = No, already has job 2 = No, temporarily ill 3 = No, other reasons (in school, etc) 4 = Yes, could have taken a job
C4	Citizen	Citizenship (Note: All in sample are U.S. Citizens) 0 = Born in U.S. 1 = Born in Puerto Rico, Guam, and Outlying 2 = Born abroad of American Parents 3 = U.S. citizen by naturalization
C5	English	Ability to speak English 0 = n/a less than 5 yrs old 1 = Very well 2 = Well 3 = Not well 4 = Not at all
C6	Fertil	Number of children ever born 0 = n/a less than 15 yrs/male 1 = No child 2 = 1 child 3 = 2 children . . . 12 = 11 children 13 = 12 or more children
C7	Hours89	Usual hours worked per week during 1989 (number of hours, except as noted) 0 = n/a under 16 yrs/ did not work in 1989 99 = 99 or more usual hours
C8	Hours	Hours worked last week (number of hours, except as noted) 0 = n/a under 16 yrs/did not work 99 = 99 or more hours last week
C9	Income1	Wage or salary income, 1989 (dollars, except as noted) 0 = n/a under 16 yrs/none
C10	Lang1	Language other than English at home 0 = n/a under 5 yrs old

		1 = Yes, speaks other language
		2 = No, speaks only English
C11	Looking	Looking for work
		0 = n/a or less than 16 yrs old/at work
		1 = Yes
		2 = No
C12	Marital	Marital status
		0 = Now married, (excluding separated)
		1 = Widowed
		2 = Divorced
		3 = Separated
		4 = Never married or under 15 yrs old
C13	Trans	Means of transportation to work
		0 = n/a, not a worker or in the labor force
		1 = Car, truck, or van
		2 = Bus or trolley bus
		3 = Streetcar or trolley car
		4 = Subway or elevated
		5 = Railroad
		6 = Ferryboat
		7 = Taxicab
		8 = Motorcycle
		9 = Bicycle
		10 = Walked
		11 = Worked at home
		12 = Other method
C14	PWGT1	{not used}
C15	Rearning	Total personal earnings (dollars; 0 = n/a or none)
C16	Riders	Vehicle occupancy on way to work
		0 = n/a
		1 through 6 = number of people
		7 = 7 to 9 people
		8 = 10 or more people
C17	SERIALNO	{not used}
C18	Sex	Sex
		0 = Male
		1 = Female
C19	TravTime	Travel time to work (number of minutes, except as noted)
		0 = n/a; not a worker or worked at home
		1 = 99 minutes or more
C20	YearSch	Years of school completed
		0 = n/a less than 3 yrs old
		1 = No school completed
		2 = Nursery school
		3 = Kindergarten
		4 = 1st, 2nd, 3rd, or 4th grade

5 = 5th, 6th, 7th, or 8th grade
6 = 9th grade
7 = 10th grade
8 = 11th grade
9 = 12th grade, no diploma
10 = HS graduate, diploma or G.E.D.
11 = Some college, but no degree
12 = Associate Degree, occupational
13 = Associate Degree, Academic
14 = Bachelors Degree
15 = Masters Degree
16 = Professional Degree
17 = Doctorate Degree

CHOLEST.MTW

Measurements of blood cholesterol were taken in heart attack patients at several-day intervals following the heart attack. These are compared with measurements from a control group of patients. N = 28 for experimental group; N = 30 for control group. Source: Minitab, Inc. Used with permission.

NOTE: In first 3 columns, each row represents one patient. Observations in column 4 are independent of the other columns.

Col.	Var.	Description
C1	2-Day	Cholesterol reading 2 days after attack
C2	4-Day	Cholesterol reading 4 days after attack
C3	14-Day	Cholesterol reading 14 days after attack
C4	Control	Cholesterol reading for control group patient

COLLEGES.MTW

Each year, *U.S. News and World Report* magazine surveys colleges and universities in the United States. The 1994 dataset formed the basis of the 1995 Data Analysis Exposition, sponsored by the American Statistical Association, for undergraduates to devise innovative ways to display data from the survey. This file contains several variables from that dataset. N = 1302. Source: *U.S. News and World Report*, via the *Journal of Statistics Education*. http://www.stat.ncsu.edu/info/jse. Used with permission. NOTE: Schools are listed alphabetically by state.

Col.	Var.	Description
C1	ID	Unique idendifying number
C2	Name	Name of school
C3	State	State in which school is located
C4	PubPvt	Public or private school (1 = public, 2 = private)

C5	MathSAT	Avg Math SAT score
C6	VerbSAT	Avg. Verbal SAT score
C7	CombSAT	Avg. combined SAT score
C8	MeanACT	Average ACT score
C9	MSATQ1	First quartile, Math SAT score
C10	MSATQ3	Third quartile, Math SAT score
C11	VSATQ1	First quartile, Verbal SAT score
C12	VSATQ3	Third quartile, Verbal SAT score
C13	ACTQ1	First quartile, ACT score
C14	ACTQ3	Third quartile, ACT score
C15	AppsRec	Number of applications received
C16	AppsAcc	Number of applications accepted
C17	NewEnrol	Number of new students enrolled
C18	Top10	Pct. of new students from top 10% of their HS class
C19	Top25	Pct. of new students from top 25% of their HS class
C20	FTUnder	Number of full-time undergraduates
C21	PTUnder	Number of part-time undergraduates
C22	Tuit_In	In-state tuition
C23	Tuit_Out	Out-of-state tuition
C24	RmBoard	Room and board costs
C25	FacPhD	Pct of faculty with Ph.D.'s
C26	FacTerm	Pct of faculty with terminal degrees
C27	SFRatio	Student-to-faculty ratio
C28	AlumCont	Pct. of alumni who donate
C29	InstperS	Instructional expenditure per student
C30	GradRate	Pct of students who graduate within 4 yrs

EUROPEC.MTW

This file is extracted from the Penn World Table data. Data values are real annual consumption as a percentage of annual Gross Domestic Product. N = 31. Soure: See **PAWORLD.MTW** (page 248).

Col.	Var.	Description
C1	Year	Observation year (1960–1992)
C2	BELGIUM	Real consumption % of GDP, Belgium
C3	DENMARK	Same, for Denmark
C4	FINLAND	Same, for Finland
C5	FRANCE	Same, for France
C6	GERMANYW	Same, for W.Germany
C7	GREECE	Same, for Greece
C8	IRELAND	Same, for Ireland
C9	ITALY	Same, for Italy
C10	NETHERLANDS	Same, for Netherlands
C11	NORWAY	Same, for Norway
C12	PORTUGAL	Same, for Portugal

C13	SPAIN	Same, for Spain
C14	SWEDEN	Same, for Sweden
C15	TURKEY	Same, for Turkey
C16	U.K.	Same, for United Kingdom

EUROPEY.MTW

This file is extracted from the Penn World Table data. Data values are real per capita GDP relative to GDP in the United States (%; U.S. = 100%). N = 31. Source: See **PAWORLD.MTW** (page 248).

Col.	Var.	Description
C1	Year	Observation year (1960–1992)
C2	BELGIUM	Real per capita GDP as % of U.S., Belgium
C3	DENMARK	Same, for Denmark
C4	FINLAND	Same, for Finland
C5	FRANCE	Same, for France
C6	GERMANYW	Same, for W.Germany
C7	GREECE	Same, for Greece
C8	IRELAND	Same, for Ireland
C9	ITALY	Same, for Italy
C10	NETHERLANDS	Same, for Netherlands
C11	NORWAY	Same, for Norway
C12	PORTUGAL	Same, for Portugal
C13	SPAIN	Same, for Spain
C14	SWEDEN	Same, for Sweden
C15	TURKEY	Same, for Turkey
C16	U.K.	Same, for United Kingdom

EXIMPORT.MTW

Current dollar value of selected US exports and imports, monthly, Jan. 1948–Mar. 1996. All numbers are millions of dollars. N = 624. Source: *Survey of Current Business.*

Col.	Var.	Description
C1	Date	Month and year
C2	A0M602	Exports, excluding military aid shipments
C3	A0M604	Exports of domestic agricultural products.
C4	A0M606	Exports of non-electrical machinery
C5	A0M612	General Imports
C6	A0M614	Imports of Petroleum and Petroleum products
C7	A0M616	Imports of automobile and parts

F500.MTW

Selected data about the 1996 Fortune 500 companies. N = 596. Source: Time, Inc. http://pathfinder.com/fortune. © 1996 Time Inc. All rights reserved. Used with permission.

Col.	Var.	Description
C1	RRank95	Revenue ranking, 1995
C2	RRank94	Revenue ranking, 1994
C3	CoName	Company name
C4	Revenue	Revenue, 1995 (millions of dollars)
C5	RevChg	Pct. change in revenue, 1994–95
C6	Profit	Profits (millions of dollars)
C7	ProfChg	Pct. change in profit, 1994–95
C8	Assets	Assets (millions of dollars)
C9	Equity	Total Stockholders Equity (millions of dollars)
C10	MktVal	Market Value 3/15/96 (millions of dollars)
C11	PrSale	Profits as a pct. of sales
C12	PrAsst	Profits as a pct. of assets
C13	PrEq	Profits as a pct. of Stockholders' Equity
C14	EPS95	Earnings per share, 1995 (dollars)
C15	EPSChg	Pct. change in EPS, 1994–95
C16	Grow8595	Annual rate of growth in EPS, 1985-95 (%)
C17	ROI95	Return to Investors, 1995 (%)
C18	ROI8595a	Annual rate of ROI, 1985-95 (%)
C19	Employ	Number of employees
C20	Indust	Primary industry
C21	State	Postal code for state of headquarters

FALCON.MTW

This worksheet contains the results of a study investigating the residual effects of DDT among falcons. DDT is a pesticide which was banned in the United States due to its long-lasting detrimental effects on bird populations. Years later, residues of DDT were still present in birds. Each row represents one bird sampled. N=27. Source: Minitab, Inc.

Col.	Var.	Description
C1	DDT	The measurement of DDT
C2	Site	Code representing the nesting site of the bird 1 = US, 2 = Canada, 3 = Arctic Region
C3	Age	Code representing the age of the bird 1 = young, 2 = middle-aged, 3 = old

GALILEO.MTW

Galileo's experiments with gravity included his observation of a ball rolling down an inclined plane. In one experiment, the ball was released from various points along a ramp. In a second experiment, a horizontal "shelf" was attached to the lower end of the ramp. In each experiment, he recorded the initial release height of the ball, and the total horizontal distance which the ball traveled before coming to rest. All units are *punti* (points), as recorded by Galileo. Sources: *Galileo at Work,* by Stillman Drake. Copyright © 1978, University of Chicago Press. Used with permission. Also see "Teaching Statistics with Data of Historic Significance", by Dickey, David A., and Arnold, J. Tim, *Journal of Statistics Education* v. 3, no. 1, 1995. Used with permission. Available via http://www.stat.ncsu.edu/info/jse.

Col.	Var.	Description
C1	DistRamp	Horizontal distance traveled, ramp experiment
C2	HtRamp	Release Height, ramp experiment
C3	DistShelf	Horizontal distance traveled, shelf experiment
C4	HtShelf	Release Height, shelf experiment

HAIRCUT.MTW

Data randomly selected from the file **STUDENT.MTW**, recording the prices paid for most recent haircut. N = 60. Source: Author.

Col.	Var.	Description
C1	Haircut	Price paid for most recent professional haircut
C2	Sex	Gender of the student (M/F)
C3	Region	Home region of the student (Rural, Suburban, Urban)

LABOR.MTW

Monthly data on employment measures from the U.S. economy, Jan. 1948–Mar. 1996. N = 624. Source: *Survey of Current Business.*

Col.	Var.	Description
C1	Date	Month and year
C2	A0M441	Civilian labor force (thous.)
C3	A0M442	Civilian employment (thous.)
C4	A0M451	Labor force participation rate, males 20 & over (pct.)
C5	A0M452	Labor force participation rate, females 20 & over (pct.)
C6	A0M453	Labor force participation rate, 16-19 years of age (pct.)
C7	A0M001	Average weekly hours, mfg. (hours)
C8	A0M021	Average weekly overtime hours, mfg. (hours)

C9	A0M005	Avg weekly initial claims, unemploy. insurance (thous.)
C10	A0M046	Index of help-wanted ads in newspapers (1987=100)
C11	A0M060	Ratio, help-wanted advertising to number unemployed
C12	A0M048	Employee hours in nonag. establishments (bil. hours)
C13	A0M042	Persons engaged in non agricultural activities (thous.)
C14	A0M041	Employees on nonagricultural payrolls (thous.)
C15	D1M963	Private nonag. employment, 1-mo. diffusion index (%)
C16	D6M963	Private nonag. employment, 6-mo. diffusion index (%)
C17	A0M040	Nonag. employees, goods-producing industries (thous.)
C18	A0M090	Ratio, civilian employment to working-age pop.(%)
C19	A0M037	Number of persons unemployed (thous.)
C20	A0M043	Civilian unemployment rate (pct.)
C21	A0M045	Average weekly insured unemployment rate (pct.)
C22	A0M091	Average duration of unemployment in weeks (weeks)
C23	A0M044	Unemployment rate, 15 weeks and over (pct.)

LONDON1.MTW

Selected hourly measurements of Carbon Monoxide concentrations in the air in West London, 1996. All measurements are parts per million (ppm). N = 325. Source: National Environmental Technology Centre. Data available at:
http://www.aeat.co.uk/netcen/aqarchive/data/autodata/1996/wl_co.csv

Col.	Var.	Description
C1	Date	Day of year
C2	1AM	ppm, CO for the hour ending 1AM GMT
C3	5AM	ppm, CO for the hour ending 5AM GMT
C4	9AM	ppm, CO for the hour ending 9AM GMT
C5	1PM	ppm, CO for the hour ending 1PM GMT
C6	5PM	ppm, CO for the hour ending 5PM GMT
C7	9PM	ppm, CO for the hour ending 9PM GMT

LONDON2.MTW

Hourly measurements of Carbon Monoxide concentrations in the air in West London, 1996. All measurements are parts per million (ppm). N = 325. Source: National Environmental Technology Centre. Data available via:
http://www.aeat.co.uk/netcen/aqarchive/data/autodata/1996/wl_co.csv

Col.	Var.	Description
C1	Date	Day of year
C2	1AM	ppm, CO for the hour ending 1AM GMT
C3	2AM	ppm, CO for the hour ending 2AM GMT
C4	3AM	ppm, CO for the hour ending 3AM GMT

C5	4AM	ppm, CO for the hour ending 4AM GMT
C6	5AM	ppm, CO for the hour ending 5AM GMT
C7	6AM	ppm, CO for the hour ending 6AM GMT
C8	7AM	ppm, CO for the hour ending 7AM GMT
C9	8AM	ppm, CO for the hour ending 8AM GMT
C10	9AM	ppm, CO for the hour ending 9AM GMT
C11	10AM	ppm, CO for the hour ending 10AM GMT
C12	11AM	ppm, CO for the hour ending 11AM GMT
C13	12NOON	ppm, CO for the hour ending 12NOON GMT
C14	1PM	ppm, CO for the hour ending 1PM GMT
C15	2PM	ppm, CO for the hour ending 2PM GMT
C16	3PM	ppm, CO for the hour ending 3PM GMT
C17	4PM	ppm, CO for the hour ending 4PM GMT
C18	5PM	ppm, CO for the hour ending 5PM GMT
C19	6PM	ppm, CO for the hour ending 6PM GMT
C20	7PM	ppm, CO for the hour ending 7PM GMT
C21	8PM	ppm, CO for the hour ending 8PM GMT
C22	9PM	ppm, CO for the hour ending 9PM GMT
C23	10PM	ppm, CO for the hour ending 10PM GMT
C24	11PM	ppm, CO for the hour ending 11PM GMT
C25	12MID	ppm, CO for the hour ending 12MIDNIGHT GMT

MARATHON.MTW

Finishing times and rankings for the Wheelchair division of the 1996 Boston Marathon. N = 81. Source: Boston Athletic Association and *Boston Globe*. http://www.boston.com/sports/marathon

Col.	Var.	Description
C1	Rank	Order of finish
C2	Name	Name of racer
C3	City	City or town of racer
C4	State	State or province of racer
C5	Country	Three-letter country code
C6	Minutes	Finish time, in minutes

MENDEL.MTW

Summary results of one genetics experiment conducted by Gregor Mendel. Tally of observed and expected frequencies of pea texture and color. N = 4. Source: Kohler, Heinz. *Statistics for Business and Economics*, 3rd ed. (New York: HarperCollins, 1994) p. 459.

Col.	Var.	Description
C1	Type	Identification of color and texture.
C2	Observed	Frequency observed

C3	Theory	Theoretical probability
C4	Expected	Frequency Expected

MFT.MTW

This worksheet holds scores of students on a Major Field Test, as well as their GPA's and SAT verbal and math scores. N = 137. Source: Prof. Roger Denome, Stonehill College. Used with permission.

Col.	Var.	Description
C1	TOTAL	Total score on the Major Field Test
C2	SUB1	Score on Part 1
C3	SUB2	Score on Part 2
C4	SUB3	Score on Part 3
C5	SUB4	Score on Part 4
C6	GPA	Student's college GPA at time of exam
C7	Verb	Verbal SAT score
C8	Math	Math SAT score
C9	GPAQ	Quartile in which student's GPA falls in sample
C10	VerbQ	Quartile in which student's Verbal SAT falls in sample
C11	MathQ	Quartile in which student's Math SAT falls in sample

NIELSEN.MTW

A.C. Nielsen television ratings for the top 20 shows, as measured during the week of September 14, 1997.N = 20. Source: A. C. Nielsen Co.

Col.	Var.	Description
C1	Rank	Ranking of the show (1 through 20)
C2	Show	Title of the program
C3	Network	Abbreviation of the broadcast network
C4	Rating	Rating score for the program that week

OUTPUT.MTW

Monthly data on output, production and capacity utilization measures from the US economy, Jan. 1948–Mar. 1996.N = 624. Source: *Survey of Current Business.*

Col.	Var.	Description
C1	Date	Month and year
C2	A0M047	Index of industrial production (1987=100)
C3	A0M073	Indust. production, durable manufactures (1987=100)
C4	A0M074	Indust. production, nondurable manuf. (1987=100)
C5	A0M075	Industrial production, consumer goods (1987=100)

C6	A0M124	Capacity utilization rate, total industry (pct.)
C7	A0M082	Capacity utilization rate, manufacturing (pct.)

PAWORLD.MTW

The Penn World Table (Mark 5.6) was constructed by Robert Summers and Alan Heston of the University of Pennsylvania for an article in the May 1991 *Quarterly Journal of Economics*. The main dataset is massive, containing demographic and economic data about virtually every country in the world from 1950 to 1992. This dataset represents selected variables and a stratified random sample of 42 countries from around the world, for the period 1960–1992. N = 1386. Source: http://cansim.epas.utoronto.ca:5680/pwt/. Used with permission of Professor Heston.
NOTE: Countries are sorted alphabetically within continent.

Col.	Var.	Description
C1	ID	Numeric Country code
C2	COUNTRY	Name of Country
C3	YEAR	Observation year (1960–1992)
C4	POP	Population in '000's
C5	RGDPCH	Per capita Real GDP; Chain Index, 1985 international prices
C6	C	Real Consumption % of GDP
C7	I	Real Investment % of GDP
C8	G	Real Government expenditures % of GDP
C9	Y	Real per capita GDP relative to U.S. (%;US = 100)
C10	CGDP	Real GDP per capita, current international prices
C11	XR	Exchange rate with US dollar
C12	RGDPEA	Real GDP per equivalent adult
C13	RGDPW	Real GDP per worker
C14	OPEN	"Openness" = (Exports + Imports)/Nominal GDP

PENNIES.MTW

Students in a class each flip 10 coins repeatedly until they have done so about 30 times. They record the number of heads in each of the repetitions. N = 56. Source: Professor Roger Denome, Stonehill College. Used with permission.

Col.	Var.	Description
C1	0 heads	Number of times (out of 30) student observed 0 heads
C2	1 head	Number of times (out of 30) student observed 1 head
C3	2 heads	Number of times (out of 30) student observed 2 heads
C4	3 heads	Number of times (out of 30) student observed 3 heads

C5	4 heads	Number of times (out of 30) student observed 4 heads
C6	5 heads	Number of times (out of 30) student observed 5 heads
C7	6 heads	Number of times (out of 30) student observed 6 heads
C8	7 heads	Number of times (out of 30) student observed 7 heads
C9	8 heads	Number of times (out of 30) student observed 8 heads
C10	9 heads	Number of times (out of 30) student observed 9 heads
C11	10 heads	Number of times (out of 30) student observed 10 heads

SALEM.MTW

Taxes paid, political factions, and status during the witchcraft trials of people living in the Salem Village parish, 1690–1692. N = 100. Source: Boyer, Paul, and Nissenbaum, Stephen, *Salem Village Witchcraft: A Documentary Record of Local Conflict in Colonial New England.* Copyright © 1993, Northeastern University Press. Used with permission.

Col.	Var.	Description
C1	Last	Last name
C2	First	First name
C3	Tax	Amount of tax paid, in pounds, 1689-90
C4	ProParris	(0-1) indicator variable identifying persons who supported Rev. Parris in 1695 records (1 = supporter)
C5	Accuser	(0-1) indicator variable identifying accusers and their families (1 = accuser)
C6	Defend	(0-1) indicator variable identifying accused witches and their defenders (1 = defender)

SLAVDIET.MTW

Per capita food consumption of slaves in 1860 compared with the per capita food consumption of the entire population, 1879. N = 12. Source: Fogel, Robert William, and Engerman, Stanley L., *Time on the Cross: Evidence and Methods—A Supplement.* Copyright © 1974, Little, Brown and Company. Used with permission.

Col.	Var.	Description
C1	Food	Food product
C2	Type	Food group (e.g., meat, grain, dairy, etc.)
C3	Slavlb	Per capita lbs consumed by slaves in 1860
C4	Slavcal	Per capita calories per day for slaves in 1860
C5	Poplb	Per capita lbs consumed by general population, 1879
C6	Popcal	Per capita calories per day, general population, 1879

SLEEP.MTW

Data describing sleep habits, size, and other attributes of mammals. N = 62. Source: Allison, T. and Cicchetti, D., "Sleep in Mammals: Ecological and Constitutional Correlates," *Science* ,v. 194, Nov. 12, pp. 732–734, 1976. Used with authors' permission. Data from http://lib.stat.cmu.edu/datasets/sleep, contributed by Prof. Roger Johnson, South Dakota School of Mines and Technology.

Col.	Var.	Description
C1	Species	Name of mammalian species
C2	Weight	Body weight, kg.
C3	Brain	Brain weight, grams
C4	Sleepnon	Non-dreaming sleep (hrs/day)
C5	Sleepdr	Dreaming sleep (hrs/day)
C6	Sleep	Total of C4 + C5
C7	LifeSpan	Maximum life span (years)
C8	Gestat	Gestation time (days)
C9	Predat	Predation index (1-5, from least to most likely to be preyed upon
C10	Exposure	Sleep exposure index (1-5;1 = sleeps in well-protected den, 5 = highly exposed)
C11	Danger	Overall danger index (1-5: least to most)

STATES.MTW

Data concerning population, income, and transportation in the 50 states of the US, plus the District of Columbia. N = 51.Source: Highway Statistics On-Line, 1990.

Col.	Var.	Description
C1	State	Name of State
C2	Pay93	Mean wages in 1993 (current dollars)
C3	Pay94	Mean wages in 1994 (current dollars)
C4	Chg9394	% change in wages, 1993 to 1994
C5	Pop	Population, 1994
C6	Area	Land area of the state, square miles
C7	Density	Population per square mile
C8	Ins92	Mean auto insurance premium, 1992
C9	Ins93	Mean auto insurance premium, 1993
C10	Ins94	Mean auto insurance premium, 1994
C11	CarsIns	Number of cars insured, 1994
C12	Regist	Number of cars registered, 1994
C13	RdMiles	Number of miles of road in the state
C14	Mileage	Avg. number of miles driven by drivers in the state

C15	FIA	Fatal Injury Accidents
C16	AccFat	Number of fatalities in auto accidents
C17	BAC	Blood Alcohol Content threshold
C18	MaleDr	Number of male drivers licensed, 1994
C19	FemDr	Number of female drivers licensed, 1994
C20	TotDriv	Total number of licensed drivers, 1994

STUDENT.MTW

This file contains results of a first-day-of-class survey of Business Statistics students at Stonehill College. All students in the sample are full-time day students. N = 219. Source: Author.

Col.	Var.	Description
C1	ID	Identifier Code
C2	Gender	F = Female, M = Male
C3	Ht	Height, in inches
C4	Wt	Weight, in pounds
C5	DOW	Day of Week on which your birthday falls this year
C6	Left	Hand you write with (0 = Right, 1= Left)
C7	Eyes	Eye Color (Blue, Brown, Green, Hazel)
C8	Maj	Major field of study (ACC = accounting, FIN = finance, MGT = management, MKT = marketing, OTH = other)
C9	First	Was Stonehill your first choice school?(Y/N)
C10	Res	Resident (R) or Commuter Student (C)
C11	WorkHr	Hours per week at a paid job
C12	GPA	Current cumulative GPA in college
C13	OwnCar	Car ownership (Y/N)
C14	Home	Miles between your home and school (est.)
C15	Region	Is your hometown rural (R), suburban (S), or urban (U)
C16	Drive	"How do you rate yourself as a driver?" (Above Average, Average, Below Average)
C17	Belt	Frequency of seat belt usage (Never, Sometimes, Usually, Always)
C18	Acc	Number of auto accidents involved in within past 2 yrs
C19	Sibling	Number of siblings
C20	Cigs	Smoked a cigarette in past month? (Y/N)
C21	Haircut	Price paid for most recent professional haircut
C22	Dog	Own a dog?
C23	Travel	Ever traveled outside of USA?
C24	Zap	Personally know someone hit by lightening(1 = yes)
C25	Beers	Number of beers consumed on Labor Day
C26	Female	Sex (1 = Female, 0 = Male)

SWIMMER1.MTW

This file contains individual-event race times for a high school swim team. Each swimmer's time was recorded in two "heats" (trials) of each event in which he or she competed. Times are in seconds. Each observation represents one swimmer in one heat of one event. N = 272. Source: Brian Carver. Used with permission.

Col.	Var.	Description
C1	Swimmer	ID code for each swimmer
C2	Gender	Gender (F/M)
C3	Heat	First or second trial (1/2)
C4	Event	Identifier of swimming event timed (length in Meters plus event code: Freestyle, Breast, Back)
C5	EventRep	Concatenation of C4 + C3
C6	Time	Recorded time to complete the event

SWIMMER2.MTW

This file contains individual-event race times for a high school swim team. Each swimmer's time was recorded in two "heats" (trials) of each event in which he or she competed. Times are in seconds. Each row contains all results for each of 72 swimmers. N = 72. Source: Brian Carver. Used with permission.

Col.	Var.	Description
C1	Swimmer	ID code for each swimmer
C2	Gender	Gender (F/M)
C3	Events	Number of different events recorded for the swimmer
C4	50mEvents	Number of 50-meter events for this swimmer
C5	100Free_1	Time in 100-meter freestyle (1st heat)
C6	100Free_2	Time in 100-meter freestyle (2nd heat)
C7	200Free_1	Time in 200-meter freestyle (1st heat)
C8	200Free_2	Time in 200-meter freestyle (2nd heat)
C9	50Back_1	Time in 50-meter backstroke (1st heat)
C10	50Back_2	Time in 50-meter backstroke (2nd heat)
C11	50Breast_1	Time in 50-meter breaststroke (1st heat)
C12	50Breast_2	Time in 50-meter breaststroke (2nd heat)
C13	50Free_1	Time in 50-meter freestyle (1st heat)
C14	50Free_2	Time in 50-meter freestyle (2nd heat)

US.MTW

Time series data about the U.S. Economy during the period from 1965–1996.N = 32. Sources: *Statistical Abstract of the United States*, *Economic Report of the President*, various years.

Col.	Var.	Description
C1	Year	Observation Year
C2	Pop	Population of the U.S. for the year (000's)
C3	Employ	Aggregate Civilian Employment (000's)
C4	Unemprt	Unemployment Rate (%)
C5	GNP	Gross National Product (billions of current $)
C6	GDP	Gross Domestic Product (billions of current $)
C7	PersCon	Aggregate Personal Consumption (billions)
C8	PersInc	Aggregate Personal Income (billions)
C9	PersSav	Gross Personal Savings (billions)
C10	DefGDP	GDP Price Deflator (1987 = 100)
C11	DefPC	Personal Income Deflator (1987 = 100)
C12	CPI	Consumer Price Index (1982–84 = 100)
C13	M1	Money supply (billions)
C14	DOW	Dow-Jones 30 Industrials Stock Avg.
C15	Starts	Housing Starts (000's)
C16	Sellprc	Median selling price of a new home (current $)
C17	ValNH	Value of new housing put in place (current mil. $)
C18	NHMort	New home mortgage interest rate
C19	PPIConst	Produce Price Index for construction materials
C20	Cars	Cars in use (millions)
C21	MortDebt	Aggregate mortgage debt (billions, current)
C22	Exports	Total exports of goods and services (bil.., current)
C23	Imports	Total Imports of goods and services (bil., current)
C24	FedRecpt	Total Federal receipts (billions, current $)
C25	FedOut	Total Federal outlays (billions, current $)
C26	FedInt	Interest paid on Federal debt (billions, current)

UTILITY.MTW

Time series data about household usage of natural gas and electricity over a period of years in the author's home. N = 81.Source: Carver, R. "What Does it Take to Heat a New Room?" *Journal of Statistics Education,* v. 6, no. 1, 1998. Used with permission. Available at http://www.stat.ncsu.edu/info/jse.

Col.	Var.	Description
C1	Month	Month and year of observation
C2	Days	Days in the month

C3	MeanTemp	Mean temperature in Boston for month
C4	GaspDay	Average number of "therms" of natural gas consumed
C5	Therms	Total therms used during month
C6	GasDays	Number of billing days in month (gas)
C7	KWH	Kilowatt-hours of electricity consumed
C8	KWHpDay	Kilowatt-hours of electricity (avg) per day
C9	ElecDays	Number of billing days in month (electric)
C10	Est	Electricity bill based on actual reading or estimate (0 = actual, 1 = estimate)
C11	HDD	Heating degree-days in the month[2]
C12	CDD	Cooling degree days
C13	NewRoom	Dummy variable indicating when the house was enlarged by one room.

WATER.MTW

Data concerning fresh-water consumption in 221 water regions throughout the United States, for the years 1985 (columns 2 through 17) and 1990 (columns 18 through 33). All consumption figures are in millions of gallons per day, unless otherwise noted. N = 221. Source: U.S. Geological Survey. http://water.usgs.gov/public/watuse/wudata.html

Col.	Var.	Description
C1	area	Region identifier code
C2	po-total85	Total population of area, thousands
C3	ps-wtofr85	Total fresh water withdrawals
C4	ps-prcap85	Per capita water use, gallons per day
C5	co-cuse85	Commercial consumptive use
C6	do-cuse85	Domestic consumptive use
C7	in-cufr85	Industrial fresh water consumptive use
C8	pt-cufr85	Thermoelectric power fresh water consumptive use
C9	pf-cufr85	Thermoelectric power (fossil fuel) fresh water consumptive use
C10	pg-cufr85	Thermoelectric power (geothermal) fresh water consumptive use
C11	pn-cufr85	Thermoelectric power (nuclear) fresh water consumptive use
C12	mi-cufr85	Mining fresh water consumptive use
C13	lv-cuse85	Livestock fresh water consumptive use
C14	ir-convy85	Irrigation conveyance losses
C15	ir-cuse85	Irrigation fresh water consumptive use

[2] A 'degree-day' equals the sum of daily mean temperature deviations from 65° F. For heating degree days, only days below 65° F are counted. For cooling degree days, only days warmer than 65° F are counted.

C16	to-frtot85	Total fresh water use (all kinds combined)
C17	to-cufr85	Total fresh water consumptive use
C18	po-total90	Total population of area, thousands
C19	ps-wtofr90	Total fresh water withdrawals
C20	ps-prcap90	Per capita water use, gallons per day
C21	co-cuse90	Commercial consumptive use
C22	do-cuse90	Domestic consumptive use
C23	in-cufr90	Industrial fresh water consumptive use
C24	pt-cufr90	Thermoelectric power fresh water consumptive use
C25	pf-cufr90	Thermoelectric power (fossil fuel) fresh water consumptive use
C26	pg-cufr90	Thermoelectric power (geothermal) fresh water consumptive use
C27	pn-cufr90	Thermoelectric power (nuclear) fresh water consumptive use
C28	mi-cufr90	Mining fresh water consumptive use
C29	lv-cuse90	Livestock fresh water consumptive use
C30	ir-convy90	Irrigation conveyance losses
C31	ir-cuse90	Irrigation fresh water consumptive use
C32	to-frtot90	Total fresh water use (all kinds combined)
C33	to-cufr90	Total fresh water consumptive use
C34	percentcu	Consumptive use at % of total use, 1985
C35	frchg	Change in fresh-water use, 1985 to 1990
C36	frchgpc	% change in fresh-water use, 1985 to 1990

WEB.MTW

Results of 20 sets of 20 trials using the Random Yahoo! Link. "Problem" defined as encountering an error message or message indicating that the referenced site had moved. N = 20. Source: Yahoo!®. Use this bookmark to activate the Random Yahoo! Link: http://random.yahoo.com/bin/ryl

Col.	Var.	Description
C1	Sample	Sample number (1–20)
C2	n	Sample size (equals 20 in all samples)
C3	Problems	Number of problems encountered in n repetitions.

WORLD90.MTW

This file is extracted from the Penn World Tables dataset described above. All data refer only to the year 1990. N = 58. Source: See **PAWORLD.MTW**, (page 248).

Col.	Var.	Description
C1	ID	Numeric Country code
C2	COUNTRY	Name of Country
C3	POP	Population in '000's
C4	RGDPCH	Per capita Real GDP, using a Chain Index, 1985 international prices
C5	C	Real Consumption % of GDP
C6	I	Real Investment % of GDP
C7	G	Real Government expenditures % of GDP
C8	Y	Real per capita GDP relative to U.S. (%;US = 100)
C9	CGDP	Real GDP per capita, current international prices
C10	XR	Exchange rate with US dollar
C11	RGDPEA	Real GDP per equivalent adult
C12	RGDPW	Real GDP per worker

Appendix B

Working with Files

Objectives

This Appendix explains several common types of files which Minitab supports and uses. Though you may not use each kind of file, it will be helpful to understand the distinctions among them. Each file type is identified by a three-character extension (like MTW or TXT) to help distinguish among them. For those just getting started with statistics and Minitab, the most useful file types are these:

Extension	File Type
MTW	Minitab Worksheet (v. 10, 11, or 12)
MTP	Minitab Portable Worksheet
TXT	Text file for Session, History, or Data
MGF	Minitab Graphics
MPJ	Minitab Project

The following sections review these types of files, and explain their use. In addition, there is a section which illustrates how you can convert data from a spreadsheet into a Minitab worksheet.

Worksheets

Throughout this manual, you have read data from Minitab worksheet files. These files have the extension MTW, and the early exercises explain how to open and save such files. These files just contain raw data (numeric, text, or date/time). In Release 12 of Minitab, additional information is embedded in MTW files, permitting the user to annotate or describe the contents of individual variables and to specify the order of values for text variables.

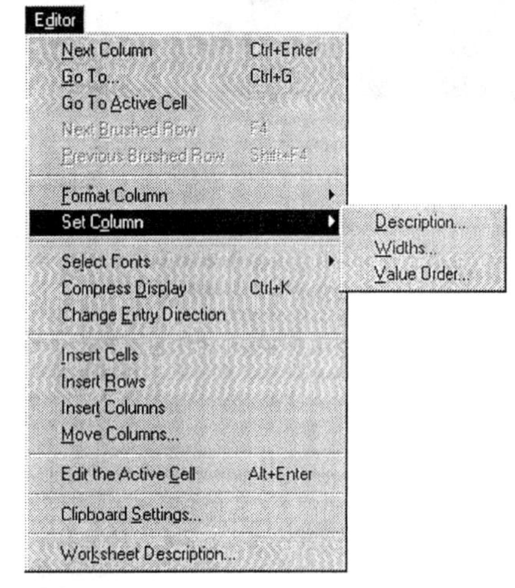

Generally, when you enter data into an active worksheet, the default settings of column format (data type, column width, and so on) are acceptable. Should you wish to customize some of these elements, you'll find relevant commands on the Editor menu. For example, as shown to the left, you can include a column description with the stored file, or specify the width of a particular column.

You can also specify the order of values for a particular variable. By default, Minitab assumes that values should be in ascending alphabetical or numerical order. However, suppose you have an ordinal variable representing size, and the three possible values are "Small, Medium, and Large." When Minitab computes a Tally or constructs a Chart for the variable, the values will be ordered as "Large, Medium, Small." With the Value Order command, you can enforce the order which is meaningful.

Earlier versions do not allow these features, which is one reason that the **Save Worksheet As** dialog lists several Minitab formats for saving data files (as shown here):

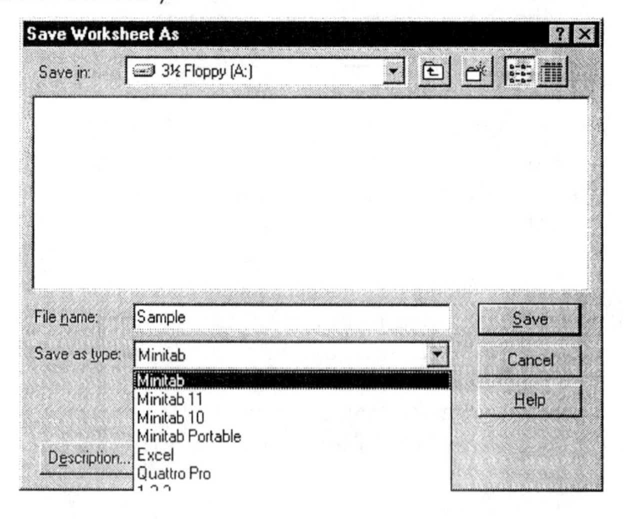

Also note that data in a worksheet can be saved as a Minitab Portable file, or in one of several popular spreadsheet or database formats. The latter options are discussed later. The "Portable" format is a version-independent format for earlier versions of Minitab. If you need to save a worksheet for other Minitab users, but aren't sure which version of Minitab they run, this is the safe choice of file formats.

Note that when you save a Worksheet file, you are *only* saving the data from a given session. If you also want to save the results of analysis, or the commands you have given, you must save a Session or History file.

Session and History Files

After doing analysis with Minitab, you may want to save a record of the work you've done, particularly if you need to complete it at a later time. That is the point of Session and History files. These files are "transcripts" of the output and command (respectively) you have given during a working session. By default, they are 'text files' meaning that they contain no formatting (boldface, etc.), but only contain words and numbers.

Each of these two file types contains everything that you see in the respective Minitab Window. The Session output is familiar enough, but the History file might appear a bit odd. As you select commands and options from various menus and dialogs, Minitab translates these choices into a command language. A History file contains all of your choices, expressed in that language. To save a Session or History file, first select either the Session or History Window, as appropriate, by clicking anywhere in the window. Here is an example, assuming the Session Window is active:

File ➤ Save Session Window as...

Make this selection to save Session output. Minitab automatically appends the TXT suffix, though you could select Rich Text Format, allowing you to retain formatting of the Session Window.

Naturally, if a History Window is active, the menu selection refers to History; otherwise, the commands are the same.

Graph Files

In addition to Session Window output, various commands create graphs. Each graph is an object in its own right, appearing in its own window. Likewise, an active graph can be saved in a file. By default, when you save a graph, Minitab stores it as a Minitab Graphics File (MGF extension); such a file can be re-opened, displayed or printed during a later Minitab session (**File ➤ Open Graph**).

Alternatively, you can save graphs in one of several graphics formats (JPG, TIF, BMP) permitting you to exchange a graph object with other application software.

Minitab Projects

New to Version 12 is the Minitab Project file (MTP) format. A project file is analogous to a binder in which you might store several related items. A Minitab Project file stores a collection of Data Worksheets, Session, History and Graphics files all in one place.

To create a Project file, you select

File ➤ Save Project By default, this command saves all open windows. To exclude an item, just close that window.

Similarly, when you open a project (**File ➤ Open Project...**), all of the saved windows and elements re-appear on the screen, allowing you to resume work right where you left off.

Converting Other Data Files into Minitab Worksheets

Often one might have data stored in a spreadsheet or database file, or want to analyze data downloaded from the Internet. Minitab can easily open many types of files. This section discusses two common scenarios; for other file types, you should consult the extensive Help files provided with Minitab. Though you do not have the data files illustrated here, try to follow these examples with your own files, as needed.

Excel spreadsheets

Suppose you have some data in a Microsoft Excel spreadsheet, and wish to read it into a Minitab worksheet. You may have created the spreadsheet at an earlier time, or downloaded it from the Internet. This example shows how to open the spreadsheet from Minitab.

First, it helps to structure the spreadsheet with variable names in the top row, and reserve each column for a variable. Though not necessary, it does simplify the task. Such a spreadsheet is shown here:

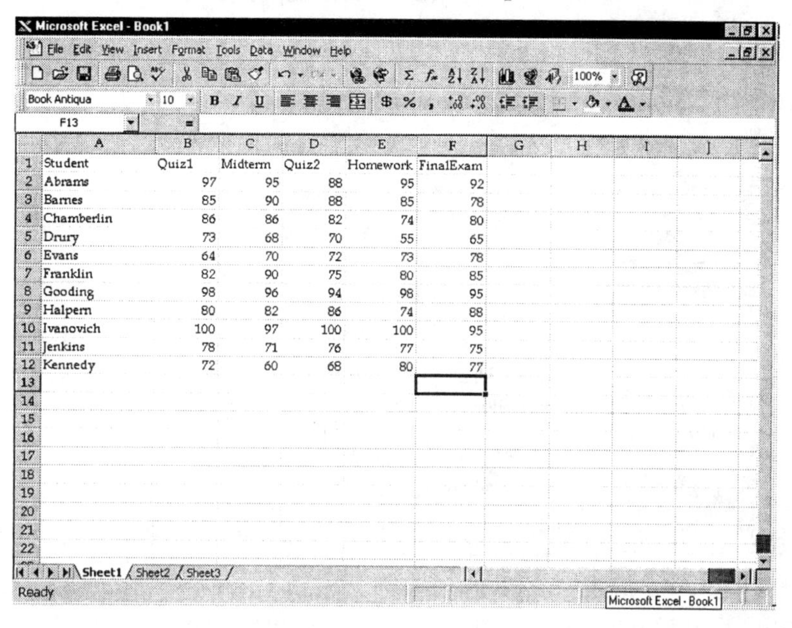

Student	Quiz1	Midterm	Quiz2	Homework	FinalExam
Abrams	97	95	88	95	92
Barnes	85	90	88	85	78
Chamberlin	86	86	82	74	80
Drury	73	68	70	55	65
Evans	64	70	72	73	78
Franklin	82	90	75	80	85
Gooding	98	96	94	98	95
Halpern	80	82	86	74	88
Ivanovich	100	97	100	100	95
Jenkins	78	71	76	77	75
Kennedy	72	60	68	80	77

Assuming the spreadsheet has been saved as an Excel (XLS) file, called **Grades**, you would proceed as follows in Minitab:

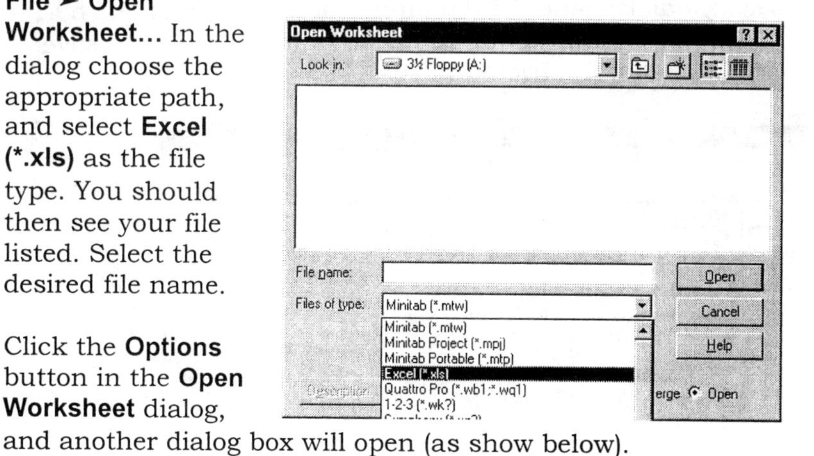

- **File ➤ Open Worksheet...** In the dialog choose the appropriate path, and select **Excel (*.xls)** as the file type. You should then see your file listed. Select the desired file name.

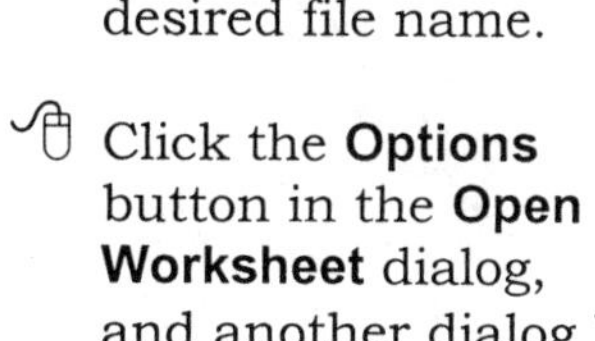

- Click the **Options** button in the **Open Worksheet** dialog, and another dialog box will open (as show below).

If information in the spreadsheet is arranged as described earlier (variables in columns, and variable names in the first row) the **Options** are unnecessary. However, if there are no variable names, or if the data begin in "lower" rows of the spreadsheet, this dialog permits you to specify where Minitab will find the data.

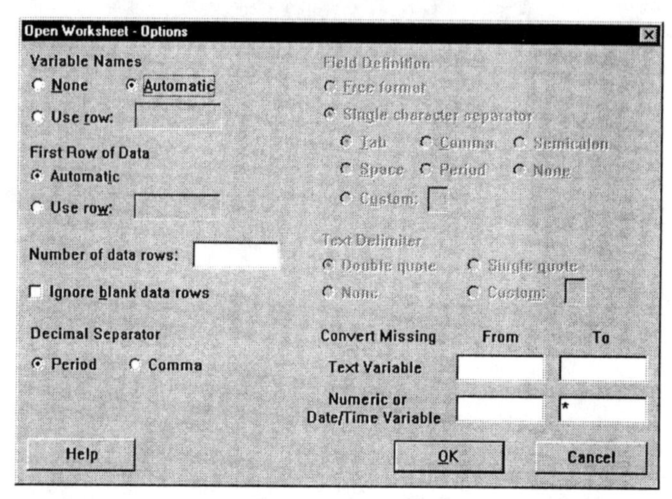

Moreover, the **Options** dialog also allows for specific instructions regarding the treatment of missing data. In this example, we have no need of the options.

Click **OK** on the **Options** dialog, and **OK** on the **Open Worksheet** dialog, and the data from the spreadsheet will be read into the Data Window. You can then analyze it and save it as a Minitab worksheet.

Grades.xls ***

	C1-T	C2	C3	C4	C5	C6
↓	Student	Quiz1	Midterm	Quiz2	Homework	FinalExam
1	Abrams	97	95	88	95	92
2	Barnes	85	90	88	85	78
3	Chamberlin	86	86	82	74	80
4	Drury	73	68	70	55	65
5	Evans	64	70	72	73	78
6	Franklin	82	90	75	80	85
7	Gooding	98	96	94	98	95
8	Halpern	80	82	86	74	88

Data in text files

Much of the data available for downloading from Internet sites is in text files, sometimes referred to as ASCII format.[1] As just described, Minitab can read data from these files, but needs to be told how the data are arranged in the file.

In these files, observations appear in rows or lines of the file. Text files generally distinguish one variable from another either by following a fixed spacing arrangement, or by using a separator or "delimiter" character between values. Thus, with fixed spacing, the first several rows of the student grade data might look like this:

```
Abrams      97    95    88    95    92
Barnes      85    90    88    85    78
Chamberlin  86    86    82    74    80
Drury       73    68    70    55    65
Evans       64    70    72    73    78
Franklin    82    90    75    80    85
```

In this arrangement, the variables occupy particular positions along each line. For instance, the student's name is within the first 11 spaces of a given line, and his/her first quiz grade is in positions 12 through 14.

Alternatively, some text files don't space the data evenly, but rather allow each value to vary in length, inserting a pre-specified character (often a comma) between the values, like this:

```
Abrams, 97, 95, 88, 95, 92
Barnes, 85, 90, 88, 85, 78
Chamberlin, 86, 86, 82, 74, 80
Drury, 73, 68, 70, 55, 65
Evans, 64, 70, 72, 73, 78
Franklin, 82, 90, 75, 80, 85
```

Logically, both of these lists of data contain all of the same information. To our eyes and minds, it is easy to distinguish that each list represents six variables. Though there is no single "best format" for a text file, it is important that we identify the format to Minitab, so that it correctly interprets what it is reading. To see why this matters, let's

---

[1] ASCII stands for American Standard Code for Information Interchange. Unlike a Minitab worksheet or other spreadsheet format, an ASCII file contains no formatting information (fonts, etc.), and only contains the characters which make up the data values.

assume that the text file contains the data evenly spaced in pre-specified positions (as in the first example).

In the **Open Worksheet** dialog, we select the **Text (*.txt)** file type. Before opening the file, we click on **Preview**, and see:

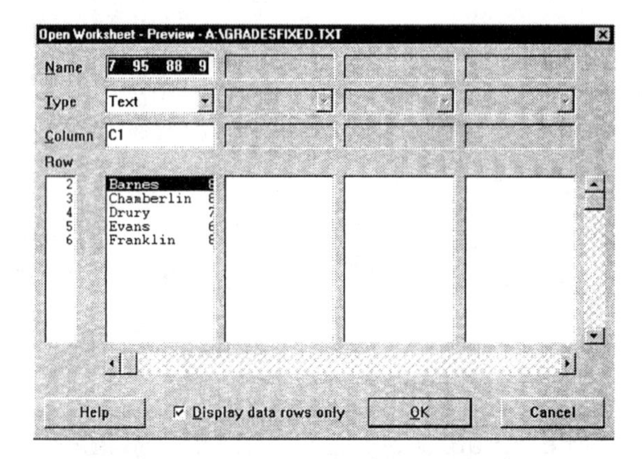

This dialog shows that all of the data will be read into Column C1, and assigned a rather odd name. Clearly, this was not what we had in mind. There are two problems here. First, Minitab is interpreting the first row of data as if it contained variable names. Second, it thinks that each line is one long text variable, rather than six individual variables.

Here is where the **Options** dialog can help. This completed dialog indicates that there are no variable names in this text file, and that the data are not separated by any single special character.

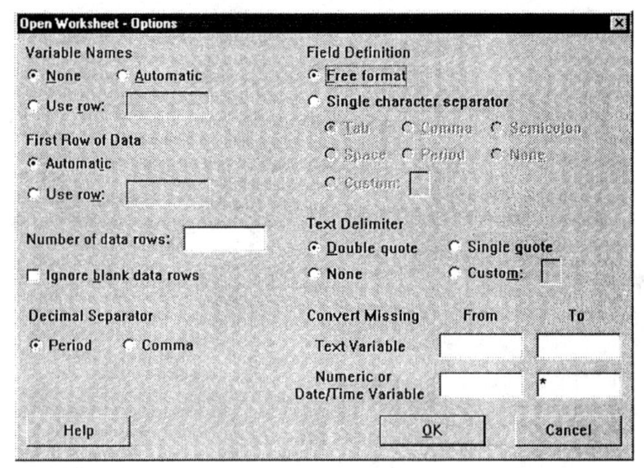

With the options as shown, a preview of the worksheet now looks like this:

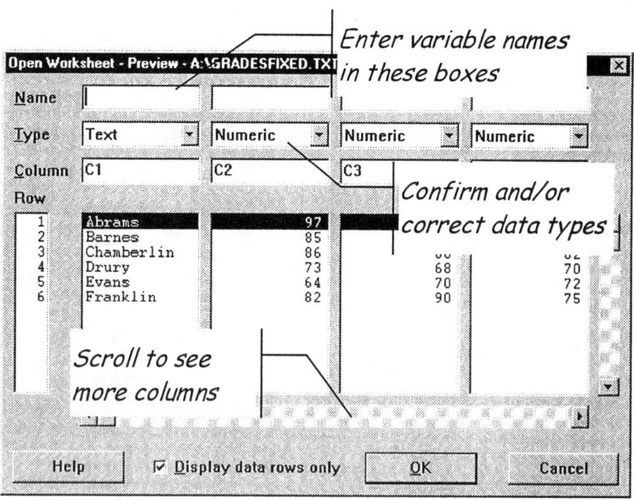

This is what we wanted. By clicking **OK** in the **Preview** and the **Open Worksheet** dialogs, the data will be read into the Data Window as before.

Though this discussion has not covered all possibilities, it does treat several common scenarios. By using the Help system available with your software, and patiently experimenting, you will be able to handle a wide range of data sources.

Appendix C

Organizing a Worksheet

Choices

Most of the datasets accompanying this book contain a single sample of subjects (states, companies, people, etc.). Within the worksheet, each row contains information about one subject, and each column represents a variable or a characteristic of the subject.

This structure represents one of three fundamental approaches to building a worksheet, and most Minitab commands anticipate (if not require) that the data be organized in this way. Nonetheless, there are times when a different approach might be necessary or desirable. The goals of this section are to help you select a method of organization for data you might collect, and to work more readily with data provided with this book or your textbook.

Consider a small class of 10 students, with the following information concerning each person:

Name	Sex	Height	Class
Amy	F	61	JR
Ben	M	65	JR
Charles	M	72	SO
Debra	F	66	SR
Elliott	M	70	JR
Francine	F	62	JR
George	M	74	SO
Hannah	F	64	SR
Ivan	M	65	SR
Jaime	F	66	JR

We'll consider three ways to represent these data in a Minitab worksheet. The first, which is most common in our worksheet files, is known as *stacked data* format. Stacked format is the most versatile of the arrangements, permitting virtually any Minitab operation.

Stacked Data

Most Minitab procedures and commands are designed to operate on stacked data. In this arrangement, each column represents a variable or attribute of one subject in the sample, and each row represents one subject. Thus, all of the observations for a given variable are 'stacked up' in a single column. The sample size (n) equals the number of rows.

Worksheet 1 ***

	C1-T	C2-T	C3	C4-T
↓	Name	Sex	Height	Class
1	Amy	F	61	JR
2	Ben	M	65	JR
3	Charles	M	72	SO
4	Debra	F	66	SR
5	Elliott	M	70	JR
6	Francine	F	62	JR
7	George	M	74	SO
8	Hannah	F	64	SR
9	Ivan	M	65	SR
10	Jaime	F	66	JR

Note that we could think of these data as representing samples from two populations (females and males) or three populations (Sophomores, Juniors, and Seniors), depending on the questions and issues under study.

When organized as stacked data, the sub-populations are identified by a variable (Sex or Class). In contrast, "unstacked data" assigns observations for a single variable to *different* columns, depending on the sub-population.

Unstacked Data

Suppose that our major interest were to compare the heights of male and female students. In that case, we might segment our data into two sub-samples, and enter it into the worksheet this way:

C5-T	C6	C7-T	C8-T	C9	C10-T
FName	FHeight	FClass	MName	MHeight	MClass
Amy	61	JR	Ben	65	JR
Debra	66	SR	Charles	72	SO
Francine	62	JR	Elliott	70	JR
Hannah	64	SR	George	74	SO
Jaime	66	JR	Ivan	65	SR

Arranged in this way, the Sex variable is omitted, since it is implied by the separation of Names, Heights, and Classes. Logically, this contains all of the information as in the stacked arrangement. Moreover, some Minitab commands (such as the Paired t-test) require unstacked data.

Summarized Data

Sometimes, we begin our work with summary results. For example, we may want to analyze a table published in a journal or news article. We don't actually have access to the original raw data, but instead have only a cross-tabulation or frequency distribution. This limits our analytical options severely, but not completely. One possible summary of our student is as follows:

C12-T	C13	C14
Year	Female	Male
SO	3	2
JR	0	2
SR	2	1

Clearly, we have lost a good deal of information here—we no longer record the names or heights of the individual students. We don't know which three females are sophomores. Nonetheless, we could still analyze this table, say, to see if gender and class were independent.

Appendix D

Working with Other Minitab Releases

Objectives

This Appendix reviews ways in which you can complete the exercises in this book using the Student Version of Minitab 12 or with Minitab Release 11 or earlier versions. If you are using Release 12, you should have little need for this appendix.

Differences Between Release 12 and Earlier Versions

Several of the exercises in this book take advantage of the new commands and capabilities of Minitab Release 12. Earlier versions of Minitab call for minor adaptations of the lab sessions. In some commands, you'll notice minor changes in the appearance of a dialog box or the name of a menu option. Other changes, though, are more substantial. Specifically, the important changes and new capabilities relevant to this book are in these commands and functions:

- Data and File Management
  - Project files, containing worksheets and other elements
  - Ability to split and subset a worksheet
  - Support for multiple open worksheets
- Descriptive Statistics
  - Dotplot
- Inferential Statistics
  - Paired (matched-sample) t-tests
  - One- and two-sample tests of proportions

Issues for Student Version 12 Users

For those using the Student Version, the major issue deals with worksheet size. Many of the datasets provided with this book exceed the limit of 5,000 worksheet cells in the Student Version. Fortunately, you can complete all of the lab sessions and problems by selecting variables for analysis.

The strategy is simple: discard the worksheet variables that are not needed for the current session. When the Student Version attempts to open a worksheet with more than 5000 cells, you will see the dialog box shown here (the example is the **Census90** worksheet):

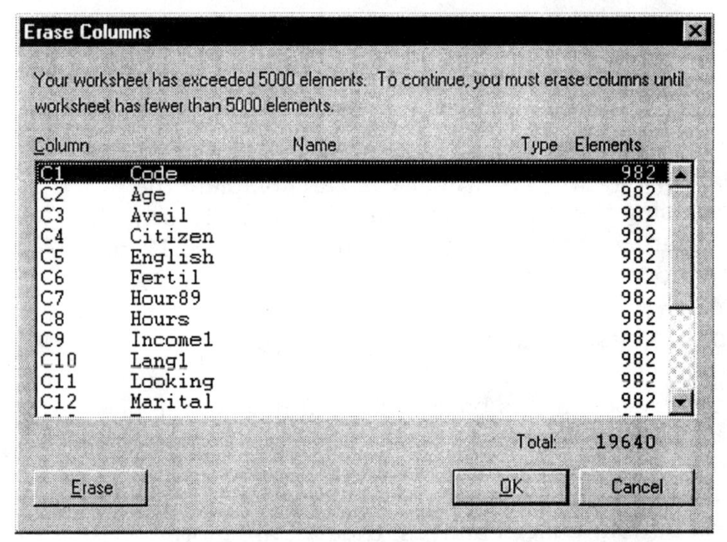

When you see this dialog, read ahead in the lab session to identify those variables which you'll need to complete the session. Then, by highlighting the names of *unneeded* variables and clicking **Erase**, Minitab will read only the remaining variables. You can repeat the process of dropping variables as often as is necessary to trim the file down to 5000 cells.

NOTE: When you "erase" variables, you are *not* deleting them from the disk file. You merely are indicating that these columns will not be read from the disk. ***However, should you wish to save the resulting worksheet, you must give it a new name, and not overwrite the original file.***

Summary of Sessions Where Differences Arise

Some of the new features, like the use of Project files or multiple open worksheets, appear in virtually every session. In earlier versions of Minitab, opening a new worksheet automatically closes the current worksheet. If you have modified a sheet, you should save it on a floppy disk (or personal directory) under an appropriate name.

Similarly, Session Window output or graphs must be saved in separate files, as needed.

As for other new features, the following table identifies those few points at which incompatibilities might arise. Suggestions for adapting to these differences appear below.

Lab Session	Split/subset	Dotplot	Paired t-test	1-sample p	2-sample p	More than 5000 cells
1						
2		☒				☒
3						☒
4						☒
5						☒
6						☒
7	☒					☒
8						☒
9						
10				☒		☒
11	☒			☒		☒
12	☒		☒			☒
13						☒
14						☒
15						
16						
17						☒
18						☒
19						
20						☒
21						☒

Workarounds for Earlier Releases

Although the following commands are new or changed in Release 12, you can perform the same task with comparable results by taking a slightly different approach. These instructions explain and illustrate methods for handling such situations.

Splitting or Subsetting a Worksheet

Splitting a Release 12 worksheet creates two or more new worksheets containing data extracted from the original. Taken together, the new sheets combined contain all of the observations from the original. *Subsetting* a worksheet creates a new worksheet containing just a subset of the data from the original sheet. Both commands isolate a portion of the data in a worksheet for analysis, a task which can be accomplished via the Copy command. With Copy, we can create a new variable in an empty worksheet column, and we can specify which rows (observations) to include or omit as Minitab selects data to copy. An example will illustrate the technique.

In the **Census90** datafile, we might want to analyze the income of single (never-married) adults who worked. In order to isolate the income variable for just those people, we must make reference to two variables: `Income1` and `Marital` (marital status). Our strategy will be to specify two 'partial' Copy operations—one to extract just the never-married individuals, and one to omit the incomes of children.

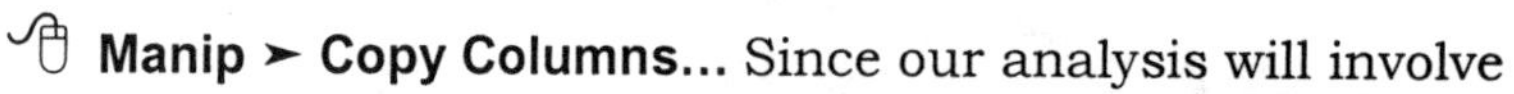

Manip ➤ Copy Columns... Since our analysis will involve income, we copy from the `Income1` data into a newly-named column (`Inc1`), as shown in this dialog. However, we only want income data from single people, so before clicking **OK**, click on **Use Rows...**.

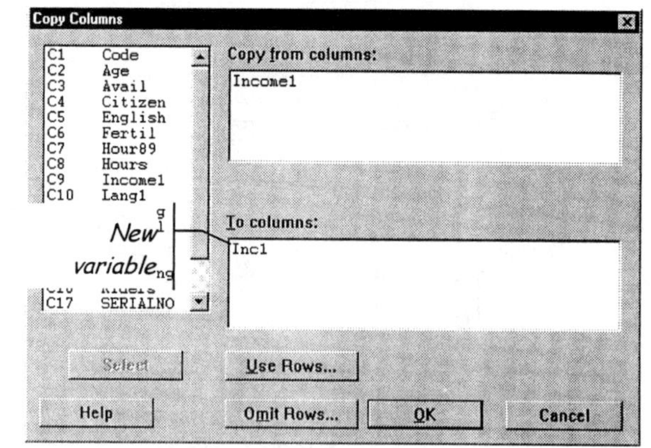

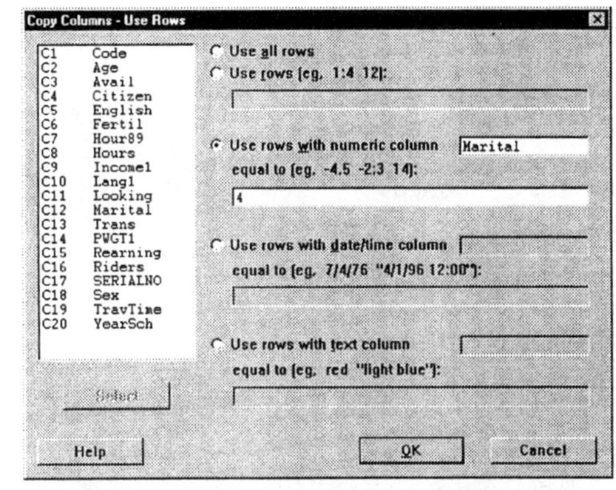

In the **Use Rows** dialog, indicate that we'll copy only those data rows for which the variable `Marital` equals 4 (indicating never married or under 15 years old). Click **OK** to return to the main copy dialog.

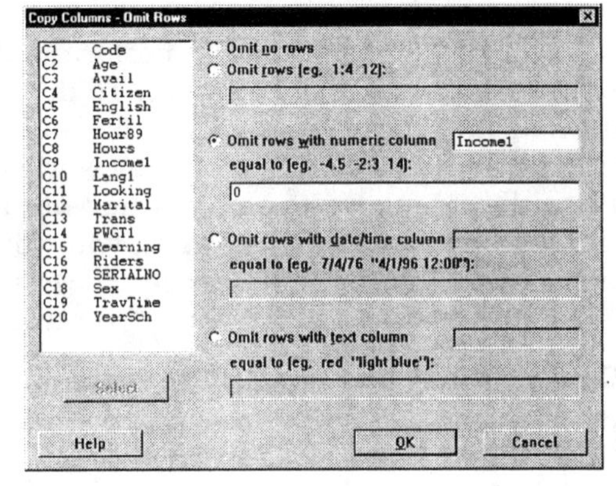

Now, click on **Omit Rows...**, to open the dialog shown here. In this case, we'll specify that we also want to skip the observations for people who were not employed; according to Appendix A, those people are represented by a value of 0 in the `Income1` variable. Click **OK**, and then **OK** in the main Copy dialog.

Dotplot

Release 12 features a high-quality graphics dotplot command, found on the **Graph** menu. In earlier releases, dotplot was a "character graph." To create a dotplot, you must select

Graph ➤ Character Graphs ➤ Dotplot...

The resulting output appears in the Session Window, and is constructed of keyboard characters, as shown in the example below. In Release 12, the dotplot would appear in its own graph window. Otherwise, the command works in the same fashion as in the current release.

Character Dotplot

```
Each dot represents 3 points
                 .
   .           . :
 : :.        ..: ::
 : :::.  ..  ::: ::  ....          :
 :::::::.::: :::::::::::. .   .  ::   :
 ::::::::::: ::::::::::::::  ::.::::..:.
.:::::::::::::::::::::::::::::::::::::::.:.  :
::::::::::::::::::::::::::::::::::::::::::::.:
+---------+---------+---------+---------+---------+---------+-------Age
0        20        40        60        80       100
```

Paired t-test

There is no explicit command to perform a matched-samples, or paired, t-test in earlier releases of Minitab. To do the test, you must have your data in unstacked form. In other words, you need your paired sample observations in two columns of your worksheet.

Next use the calculator to create a new variable, equal to the difference between the two columns. Then, you would perform a one-sample t-test on the column of differences. For example, suppose your two columns were called Before and After. You would do the following:

- **Calc ➤ Calculator** Type `Difference` in the box marked **Store result in variable:** and type `After - Before` in **Expression**. Click **OK**.

- **Stat ➤ Basic Statistics ➤ 1-Sample t...** The variable is `Difference`. Specify the appropriate alternative hypothesis, and click **OK**.

Commands Without Equivalents in Earlier Releases

Unfortunately, confidence intervals and tests concerning population proportion were not available in earlier versions of Minitab. One can certainly use the descriptive statistics and calculator features to aid in computations, but these techniques (in Sessions 10, 11, 12) are best done by hand. Note that the affected sessions contain other instructions that *are* suited for earlier releases of Minitab.

Index

Index

Index

C

D

E

F

G

H

I

J

K

L

O

P

Q

R

S

T

Y

Z